Veterans:

Rebuild America

Printed in the United States of America.
First Printing, 2014.

ISBN-10: 1941944035
ISBN-13: 978-1-941944-03-5

For group and institution discounts, please visit: www.ChadGrills.com/books

For My Wife, Stephanie.

Also By The Author

Future Proof

Collections:

Short Stories: Veterans After War

Inbox Inspiration

Does anyone want to get *more* email?

Probably not. But I think some of us want more *great* emails.

Great emails must be packed with value, inspiration, and creativity.

I send out a weekly email newsletter that aspires to deliver just that. Each one is packed with:

- Reading recommendations
- The latest articles I've written around the web
- Notifications for when my books go on sale or are free
- Answers to Q&A questions from readers
- The best videos and podcasts

You can sign up to get my newsletter for free at www.chadgrills.com

It might be the *best* email you get all week.

I know! That's a bold…maybe crazy claim. But I'm willing to bet it's true.

Best,

Chad Grills

"There are a thousand hacking at the branches of evil to one who is striking the root."
–Henry David Thoreau

Dear Reader,

If you've taken the time to even pick up this book, I thank you. Whether you're a veteran, family member, or someone who knows a veteran, I'm glad you're here.

When veterans return home, we've asked them to "reintegrate." After the extreme challenges veterans have faced, aiming to reintegrate or, "get back to normal" is impossible–and harmful.

I don't believe mediocrity is a goal worthy of our veterans' service. The biggest reason why this mentality is dangerous is because America is in trouble.

America is suffering from diseases of comfort, the national debt has exploded past $17 trillion, and our economy has stagnated – we're facing immense challenges. If we're going to rebuild America, I believe we desperately need each of our countries veterans to lead the way.

If America will survive and thrive in the future, our veterans must become the leaders, and entrepreneurs who will rebuild our country.

This book offers the pathway for those who want to use their experiences and existing skills to thrive by finding health, wealth, and wisdom in a rapidly changing world economy. *Veterans: Rebuild America* is actually made up of four smaller books. Those four smaller books focus on: health, relationships, skills, and entrepreneurship. Each one details a step-by-step process for veterans who want to become the leaders and entrepreneurs who will rebuild our country.

As a whole, *Veterans: Rebuild America* is a strategic roadmap that contains over 300+ resources that cover everything from:

- **58 Businesses** that are rebuilding America and how veterans can reverse engineer opportunities to join them
- ***BEDS***: The only four types of future-proof professions in the new economy
- How to get training for high growth, in-demand careers in as little as three months, plus over **60 connectors** to help you on your way
- ***College or Not***: How to avoid getting a worthless college degree, and how to find options outside the Post 9-11 GI Bill
- Why traditional mentorship doesn't work, and how to find the best digital mentors, plus **60 of the best teachers in the world**
- **75 business ideas** anyone can start today to begin the journey from entrepreneur to founder
- ***The Reading List of 65*** highly filtered, life changing books, plus the ***52 a year***, and ***1000 a day*** **practice**
- How-to strategically **layer skills** until we become **a monopoly of one**; the economic and logical way to find our passion
- Becoming antifragile in the new economy by mastering: **survival**, **growth**, and **signaling**
- Building opportunity generators on **LinkedIn**, **Angel List**, **Twitter**, and **Medium**
- The **CEO habit**; mindsets, experiments, and activities to find, audit, restore, and fuel healthy relationships
- The **66-day body and brain detox**
- **Competency triggers**: what they are, how to learn them, and how to instantly build rapport with successful people

- Plus the **HRSE** practice for veterans who will rebuild, restore, and lead America now and in the future

The traditional route of going to college, settling into a career and retiring happily is no longer a viable long-term option. Our world is rapidly changing.

Over 3 billion people are coming into the global economy via Smartphones. The entire developing world is learning English. These people are hungry for the opportunity Americans have, yet our culture has not fully realized what this means. The U.S. Census Bureau predicts that 68 Million people will be arriving in the United States between now and 2050. There is no way we can set ourselves, or our families, up to thrive in the future if we learn outdated skills, or enter jobs that are being replaced and outsourced by technology.

We must avoid these threats, and take advantage of all the opportunities the 21st century offers. We must each become the best, most optimal version of ourselves. This means aggressively learning and layering the right skills, finding future-proof opportunities, and allying with the industries rebuilding America. This is that path.

I'm honored to have you here.

Now let's rebuild America.

Chad Grills
Washington, DC 2014

Table of Contents

Reintegrate or Rebuild? — ***10***
What We're Facing — ***19***
The Old Way — ***33***
An Opportunity Unlike Any Other — ***41***
Functions, Culture, and Growth — ***54***
A Foundation — ***66***
Endogenous Dosing: Roots vs. Symptoms — *77*
The Mental Game: PTSD or PTG? — ***88***
The New CEO — ***107***
The Satisfied Relationship...Impossible? Or Inevitable? — ***120***
Finding, Keeping, and Building The Relationships That Matter — ***130***
Global Skills That Will Always Pay The Bills — ***140***
The Core Skills For The Future — ***151***
Mastery of Technology and Once You Go Amish — ***168***
Reverse Engineering Entry Into Fields That are Rebuilding America — ***184***
College Or Not? — ***213***
Return of the Polymath and The Passion Myth — ***235***
Survive, Grow, Signal — ***249***
From Entrepreneur to Founder — ***268***
Vale — ***303***
Terms, Appendix, and Reading List — ***309***

1

Reintegrate or Rebuild?

"The future is here. It's just not widely distributed yet."
–William Gibson

"There is so much stuff that has yet to be invented. There's so much new that's going to happen. People don't have any idea yet how impactful the Internet is going to be and that this is still Day 1 in such a big way."
–Jeff Bezos

As veterans, when we get home, we are asked to reintegrate. What a bad idea. Mediocrity and "getting back to normal" aren't goals worthy of our experiences. Veterans are some of the last Americans who have paid high prices and risked nearly everything (and sometimes given everything) during our time in the military and while deployed.

We've had our skin in the game for what we believe. Few at home do. All we have to do is make a trip out to the grocery store and we see the diseases of comfort creeping in on certain portions of the American population. The afflicted are lazy, out of shape, looking to consume, endlessly wanting more things, but unwilling to take the risks or work to get them. The idea of working for a cause or business one believes in is foreign to many.

But there are also pockets of our population working hard to rebuild America. These are the people who have been taking risks while we've been away. They've been innovating to create companies and businesses which are thriving despite an economic downturn.

This group of entrepreneurs and leaders is not only fighting for the survival of this country, but they are working on preserving and benefiting all of humanity.

Through following the path of rebuilding America economically, we'll create opportunities not only for people here at home, but also those who would otherwise turn against us abroad. More economic opportunity for more people means a safer world. Poverty and lack of social mobility are the roots of so many of the challenges we see abroad. We won't ever attack this problem effectively by meddling in the affairs of others on their own turf. We can, however, address issues of global disadvantage—which lead to conflict—as a by-product of our own wealth creation and pursuit of happiness. This boosts commerce of all kinds.

There is an old saying that, "Either goods and services cross borders, or armies will." Thankfully, much of the distribution of goods can be supercharged through the technology, manufacturing, and energy industries—three in particular which are booming or poised to boom here at home in the U.S.

Our current national economy hovers around $16.8 trillion in size while our national debt (run up by career politicians) continues to rise above $17.6 trillion. The unfunded liabilities for every taxpayer are exploding in size as well… to a staggering $1.1 million per person.

We're at a point where unless our country rebounds, we're at risk of losing everything we've created. What's more, a weak America puts the entire world at risk. A world without a strong America is truly terrifying. An America viewed as weak in the world's eyes is an America vulnerable to terrorism.

America is at a tipping point. I believe that we live in the best, most moral nation in the history of mankind. We've fed the world, clothed the world, and defended the world. We're far from perfect, but the technology and innovations we've created (and continue to create) are driving all of humanity forward. We're still a nation of big thinkers, and there is an American comeback in progress. There has never been a more dangerous time for our country, but also never a better time to be an American veteran. It might not seem like it, though—and I completely understand any cynicism.

The Old Path of Credentials Is Gone

The latest scientific research finds fewer predictions of individual high achievement using standard metrics. These researchers have continued to look for predictors of success, but our traditional tests haven't seemed to predict much. They've found that the measurements of the past—IQ, the SAT, and a college diploma—don't guaranteed high achievement or happiness.

Even more rigorous academic graduate tests like the MCAT (the test doctors take to get into medical school) have come under scrutiny. The Jensen medical test of a physician's empathy is shown to be a better predictor of how the doctor will perform in the field than the MCAT. It turns out that the standardized educational model obsessed with grades and letters after one's name is pretty worthless.

So if our formal metrics of competence aren't able to predict high levels of achievement, what can? Researchers have started to point in a new direction which may seem like common sense to many. Grit and character are the only pre-requisites for achievement that we need. A more formal name for these seemingly magical hallmarks of high-level achievers is "cognitive control." With that skill, we can easily add any others we need.

We can't lose if we develop and build cognitive control, grit, and character by using adversity to our advantage. We can't lose. However, we can't learn or build these muscles in comfort. They only come from repeated experiences of uncertainty. As veterans, we're some of the last Americans who have naturally had to develop and build our character through the relentless challenges, threats, and struggles of deployments in harm's way. When we enhance the character we've built with a highly targeted plan to build our personal health, relationships, and skills—while throwing in a dash of risk taking—we can't help but create amazing lives.

Not only have the old metrics and measurements for success like IQ, the SAT, and the college diploma myth been debunked, but so has most conventional wisdom surrounding those topics. We'll dive into the new science and research, but most importantly, we'll lay out all of these concepts as a buffet. There is no individualized prescription, just a smorgasbord where we can browse for personal solutions. Each new concept is designed to be tested, used as an experiment, and applied or discarded depending on results. These templates range from how to get into contact with decision makers, to how to reverse engineer the college-or-not choice, and also how to find a future-proof job.

Now is the time for veterans to capitalize on this opportunity. No matter what our interests, we'll see how to ally ourselves with the entrepreneurs, leaders, politicians, and venture capitalists currently taking the risks necessary to ensure that America thrives. We'll see how to contact them, how to learn the right skills to break into these industries, and how we can have a great time and rebuild our relationships while doing so.

I'll admit, some days I'm terrified considering the future of the country. It's staggering to see many people who are unwilling to take the personal and professional risks necessary to spark progress. Watching those who are looking to others to help them might be demoralizing, but those people still don't make up the majority of our country, or so this author believes. Many of these people who we view as lazy, uncaring, or unpatriotic are just demoralized. They're looking for something and someone to believe in. That's where veterans come in.

If we look at public sentiment, it's clear the American public still supports us, especially over some of our current “leaders.” Right now the general support for veterans from the public, depending on the poll, is around 97%. For politicians, Congress, and others in power, support is at a staggering 7%. The general public isn’t satisfied with Congress, but couldn’t approve more of the job we do. It’s time to use the public sentiment to our advantage.

There are around 2.8 million veterans of Iraq and Afghanistan currently in America, and over 22 million veterans total. I don’t believe any of us are ready to “reintegrate” or sink back into where we were before. We’ve had the experiences of hardship and challenge which have built our character.

It’s time to apply those here at home. We’re ready for a new challenge, one that is worthy of our experiences and us. There are no guaranteed paths to success. But there are sure ways to failure. The biggest failure I can imagine would be to let America slip into the mindset of “reintegration”— into just another country in the world. Reintegration with mediocrity isn’t the goal. Rebuilding America is a challenge worthy of our experiences.

In a little over 238 years, our country has risen from a handful of farmers fighting the British Empire to the greatest nation on earth. There is something special and unique about America. Our past is great, but our future can be even better. Rebuilding the country starts with each one of us.

It’s on us to become leaders in the American industries giving our country an opportunity to rebound on a massive scale. These opportunities lie largely in technology—the 12 largest American technology companies have revenues of $2 trillion.

If we want a free and prosperous America in the future, we must become the entrepreneurs and leaders to take the country there. We have the grit, character, and courage, which are the only real prerequisites needed for great achievement. There are other things we have to pick up, but they all can be learned as we go.

Strategy and Action

"Just as an individual's ability to delay gratification at a young age is a powerful predictor of future academic and professional achievement, discipline is also central to the long-run economic health of nations." –Peter Blair Henry

"Long-term, we must begin to build our internal strengths. It isn't just skills like computer technology. It's the old-fashioned basics of self-reliance, self-motivation, self-reinforcement, self-discipline, self-command." –Steven Pressfield

The Marshmallow Exercise

By now, we've all heard of the Marshmallow Exercise. It's named after a famous test done by researchers in the late 1960s and 1970s at Stanford.

It went like this: researchers ran a series of tests where children were incentivized to wait for a reward. For instance, a researcher would ask a child if they preferred one marshmallow now, or to wait a short period and then receive two marshmallows. Some children chose to delay their gratification just a bit longer than the rest. They waited the specified period of time, and then received two marshmallows. The children who could delay their gratification longer were found to have higher achievement later in life.

Researchers have continually found that among adults, the same behavior is observed; those who can delay gratification longer are higher achievers. This ability to delay gratification is called cognitive control.

Our own cognitive control is a skill. It can be learned, built, and mastered even later in life. Scientists have now introduced this concept as "neural plasticity." The concept of neural plasticity challenges us with the findings that we can each rewire our brain to build new skills and abilities. Top researchers and doctors around the world used the think of our brain as being "hardwired". They wrongly assumed that once we became adults our cognitive capabilities became fixed, and then gradually declined. Dr. Norman Doidge in *The Brain That Changes Itself* shows us that not only are our abilities not fixed, but also how quickly our brain responds to training. We can build new habits and form new synaptic connections after only small amounts of practice.[1] For instance, that practice doesn't always look like a science experiment. Here are a few examples of activities which build new synaptic connections to rewire our brain:

Reading and writing
Creating or building anything
Aerobic exercise
Delaying our own gratification (the cornerstone of them all)

But let's start with the most important practice, delaying gratification. Maybe we put off buying tickets for the game until after we accomplish (whatever it is we want here). This small and trivial act builds the habit of delaying gratification, one of the most important predictors of success.

The reason we bring up this simple exercise is because if we can delay our own gratification, even minutely, day by day, we build up our cognitive control. It's hard to believe that a task so simple commands such a premium on achievement, but it does.[2] From this skill, we can modify, adjust, and build all other aspects of our behavior.

The practice we need to build the neural pathways of cognitive control is a lot like exercise. Let's start by running a simple delayed gratification, "marshmallow" exercise today, in order to boost cognitive control tomorrow.

Grab a sheet of paper or feel free to write in the book. Write down three things you can't wait to do or will get a ton of satisfaction from doing. These will be our "marshmallows" or rewards.

1.
2.
3.

Now write down three things you want to accomplish. After you do them, feel free to reward yourself with any of the three rewards listed above. Add a few ideas or strategies on how you can delay the reward until you've achieved what you want. Feel free to apply this practice to ridiculously trivial things. The important part is just building this practice.

1.
2.
3.

That's it. Just by getting in the habit of postponing our rewards, we're actively building our cognitive control. We're used to credentials and grades being hailed as the prerequisites for success, but this simple act of delaying gratification is the actual foundation of everything we need for success. When we practice delaying personal gratification (briefly at first) the rewards grow in size and get sweeter.

Delaying gratification, even by just a few hours, is a practice which stretches our ability to be patient. This skill is in incredibly short supply in our modern technological era. With this mindset of stretching and delaying gratification, we're able to endure the mental agitation between acting and arriving at rewards.

Those who spend time anxiously wondering about the size, scope, or number of rewards will fail to love the process of achieving them. Those who aren't in love with the process won't endure it long enough to obtain the achievements that matter.

People who can delay gratification of their wants have a much better chance of discovering and meeting their needs. That time period when we delay gratification is crucial to discovering if our wants are real, or if they're illusory. As Richard Feynman famously said, "the first principle is that you must not fool yourself – and you are the easiest person to fool."

When we get what we want—as opposed to what we need—we find that, very shortly, we want something else. It isn't long until we've acquired a whole mess of things we never really needed, or have tricked ourselves into wanting. If our time is spent patiently discerning our needs, we build cognitive control and mastery. Now we have the chance to win gratification from meaningful achievements instead of filling our lives up with the first marshmallows that come along. Delaying gratification also helps us fight the "diseases of comfort" present in modern America, and is the foundation for the rest of the book.

2

What We're Facing

"Civilization is not inherited; it has to be learned and earned by each generation anew; if the transmission should be interrupted for one century, civilization would die, and we should be savages again." –Will and Ariel Durant

"Only an alert and knowledgeable citizenry can compel the proper meshing of the huge industrial and military machinery of defense with our peaceful methods and goals, so that security and liberty may prosper together." –President Dwight Eisenhower

Know the Enemy

The first step to overcoming the negative parts of reality is to recognize them. This chapter is about detailing what our country is facing, along with what we are facing as veterans. The only reason we present these negative realities is to better understand the enemy. So what are some of the negative realities we're facing?

The World Changed Overnight

The Internet has proliferated knowledge and made it increasingly affordable. We're currently living in a world where we can look at any creation, outcome, or result, and use the internet to discover the "how" behind it. The "why" will be up to us.

The technological revolution we're experiencing will be looked back on as larger than Gutenberg's invention of the printing press. This time period is more culturally significant than when Luther got the Bible translated from Latin into the common vernacular. The Internet is translating the alchemy used to create any result we want. Middlemen who deliver no value are being marginalized. Any process which can be standardized is becoming a commodity. We can learn more than ever before, but there has also never been more noise. We have to strategically curate, filter, and synthesize this endless stream of content into what is contextually useful for us. Never before has there been so much low cost, high quality information for those willing to set up systems to manage its consumption.

Average is No Longer Good Enough

It used to be that basic people skills and a college degree provided access to the middle class in perpetuity. But now, college is no longer a middle class insurance policy. Billions of people now have access to the Internet. Businesses are increasingly relying on remote workers who can work. Remote workers compete not only with the U.S., but also Europe and the rest of the world. College isn't enough, and the only thing that ensures we'll remain competitive is the ability to learn new things. This entrepreneurial pursuit of harnessing the Internet to become self-taught is the insurance policy of the future.

Everyone Is Still Learning English

Consider this: there are more people speaking English in India right now than in the U.K. Veterans who have deployed abroad know just how many people are desperate to learn English. We've seen all those who want to be just like Americans. I'm wondering when we will start speaking "American" and be proud of it? The world is watching us, learning English, and trying to be like us. America is still the land of innovation, opportunity, and freedom.

The World Wants Our Opportunity

The U.S. Census Bureau predicts that between now and 2050, 68 million immigrants (both legal and illegal) will arrive in the U.S.[3] Many immigrants in this flood of people will be highly skilled and well trained. They haven't spent years in the comfort of our country. They will have sacrificed to get here—as our ancestors did—and will seize opportunities where they find them. It's important for us to be prepared so we aren't blindsided by this stream of immigrants who are hungry to work.

Let's look further at just how badly immigrants want to be here. They want to be here so badly that the EB-5 visa program is skyrocketing in popularity. EB-5 is a nickname for the government's fifth employment-based visa preference. This is a government program which charges those who wish to immigrate here $500,000. This charge to enter our country must then be invested into creating 10 full-time U.S. jobs in "high unemployment or rural districts."[4] Many foreign immigrants are ready to pay $500,000 each to invest in America and get a chance to come here! Despite anything we might hear on the news, the rest of the world is hungry for the opportunity which America provides.

The Only Thing That Can Beat America?

Ourselves. We are the ones who must fight to not beat ourselves. This means we have to get the government debt under control. It means we can't overreact to the next terrorist attack and trade more civil liberties for illusions of safety. What happens when there is another event like 9/11? Are we going to spend trillions of dollars we don't have? We must strategize and act accordingly so we're not blindsided by irrationality when these potential threats arise. Every bit of government debt politicians create means mortgaging our children's future. Until we get spending and entitlement programs (unfunded liabilities) under control, we're literally selling unborn children into indentured servitude. It is on us to step up and make the type of courageous decisions necessary to rebuild the economy. The type of people we desperately need to speak up, start businesses, and generally lead the country out of this mess are those not suffering from diseases of comfort. They are those who have spent years outside of comfort. We often refer to these types as "Veterans." But what exactly are these diseases of comfort which many Americans are stricken with?

The Afflictions and Diseases of Comfort

"If you see fraud and don't say fraud you are a fraud." –Nassim Taleb

"Mercy to the guilty is cruelty to the innocent"
–Edmund Burke

Pareto's principle states that in large systems, 80% of outcomes are the result of 20% of activities. This is especially true when we talk about large systems and bureaucracies such as government, the military, or academia. When I say Americans are "suffering from diseases of comfort", I'm not being disrespectful toward the entire system. But I am recognizing that there are many "hiding out" or "gaming" these systems. These are the types who worship avoiding risk. They scheme to earn inflated salaries by transferring the risks of their activities to the general population. They fight to keep their jobs through legislation, as opposed to the merits of what they create. Once any system or bureaucracy grows too large or is weighed down by regulation, you'll inevitably get sick creatures who milk it to their advantage. I'm calling those who are sick, sick. They're crippling and inhibiting the smaller percentage of the population which is producing all the value—or in the case of veterans, taking on disproportionate amounts of risk.

Symptoms of Afflictions and Diseases of Comfort

Entitlement Programs and Mentalities: Welfare, looking to government to provide or secure jobs, "somebody should do something."

Health Epidemic: There is an obesity epidemic in the United States. The sad part is that much of this epidemic is just self-medication from past trauma. The government tried to introduce and market a food pyramid based on grain and gluten to children, and now years later we're reaping an obesity epidemic. Stay tuned for the health section of this book where we'll see how we can heal ourselves, and in turn help others heal.

Apathy, Nihilism: All the opportunity in the world, and people want to conform? Either we'll always seek to be entertained, or we'll make and create. In a land where people go from food stamps to worth a billion in a decade, why the hell is anyone lazy and bored? The last time I checked, we have companies taking humans to Mars. Who the hell has time to be lazy?

Culturally Acceptable Indulgences: Anything funded by debt financing with interest rates above inflation which is used to obtain a depreciating asset. Indulgences include most college degrees (in majors with no employment prospects), credit cards, auto loans, and mortgages. All of these indulgences are symptoms of those suffering from diseases of comfort. Culturally acceptable indulgences obtained via debt financing are a tax proportional to our ability to delay gratification.

Careers and Jobs of Buying, Selling, or Marketing these Indulgences: These are another peculiar symptom of too much comfort. These jobs and livelihoods revolve around creating, feeding, sustaining, and worshiping the glorification of indulgence over substance. They praise consuming, and turn their nose up at creating. These jobs and careers are especially insidious as mainstream media approves of them. In fact, many people use indulgences to buy their way into these salaried positions selling more indulgences. Even more people use indulgences as signals for whom to associate with. Fortunately or unfortunately, these types who are over-indulged and sick are easily identifiable in the wild.

Searching for Free Lunch: The idea that the best job is one where you do the least work and get paid the most (these jobs are often subsidized by the taxpayer).

Misunderstanding and Fear of Technology: This is prevalent where many people would rather use movies and entertainment to push widespread skepticism of technology, instead of tinkering that leads to use, creation, and meaningful dialogue.

Misuse or Underuse of Technology: The new iPhone central processing unit has around 625 times more transistors than a 1995 Pentium.[5] We have tools of learning and creation at our fingertips, yet huge numbers of people sit hunched over these devices, playing games which spam their friends with requests to play. Eventually, the lowest common denominator joins the game, makes the $99.99 in game/app purchase, and the cycle of technology abuse continues.

It's Not a Democracy: One of the most insidious habits of those stricken with the disease of comfort is their willingness to trade all our freedoms for their individual safety. The types of people most ready to do this still think we live in a democracy. The reality of our country is that our founders knew a democracy would descend into politicians dangling trinkets, promises, and indulgences in front of large voting blocks. That's why our founders started America as a democratic republic. Consider the quickest way to remember why democracies are inherently evil. Think of all the notifications we get on social networks from others spamming us to play games. Imagine that person who plays Candy Crush for hours a day. In a democracy, they have the same voting power you do. That democratic republic the founders were talking about is sounding pretty good to me.

Despite all this, the skills and traits in shortest supply are courage and creativity. Veterans have a surplus of both.

Most of the Public Has No Idea

The public misconceptions about Iraq and Afghanistan are staggering. The biggest misconception is that most of the public thinks they have been "wars." These "wars" have forced our troops to nation-build. In order to do that, a subsection of our combat arms forces has been bearing a huge amount of burden and risk. They endure crippling rules of engagement and threats of court marshal over just doing their jobs. Veterans know that if OEF or OIF had been a true "war," we would have been finished years ago.

The "wars" terminology makes speaking about these issues easier for mainstream newscasters whose skills lie in reading from a teleprompter. Asking these newscasters to properly articulate or call our nation-building politicians to task would reveal them as little more than bumbling idiots.

The academics, pundits, politicians, media, and those who have tried to socially engineer a more peaceful world through intervention often don't understand how the weight of their murky aims falls disproportionally on veterans. Our country has asked our military to nation-build for a decade, with an astounding lack of clearly defined goals. Our veterans have often found themselves enduring multiple deployments and facing death for a foreign populace uninterested in helping themselves.

No Apologies for Being the Leader

I'm not an apologist for America. We feed and protect much of the world, and we've created amazing innovations and technology which are lifting all of humanity. There are two types of leaders. The first type falls into leadership roles and grows lazy and comfortable. The second type realizes the weight and opportunity that come with leadership. This second type always seeks to become better. They don't apologize for being the best when they know they're bearing the risks associated with doing so. If America seeks to continue to lead the rest of the world, we can't waste time apologizing for being the best. We have to fight to continually ensure that America is that *shining city on the hill.*

Never Trust a Drunken Uncle in Debt

Nothing is guaranteed, only promised. In the military, we've just seen thousands of career officers who were handed "pink slips" that forced them out of the military. A "stable" career in the military is no longer a sure thing. We need to prepare ourselves for future uncertainty as our military faces over $120 billion in defense cuts.

Our benefits and pensions are not safe and secure either. Whether it's through inflation or, as the government recently tested- discuss lowering military pensions, we must prepare for change and cutbacks on the horizon. Chances are that our benefits will be there, but we can absolutely not put our full faith in them. Should we plan and do our best to get what we've been promised? Of course! But ideally, we need to create options and economic security on our own terms.

We deserve what we were promised when we entered into our contracts. But we must realize the corrupting nature of every government program as it naturally grows in size and scope, all while being administered by Uncle Sam who is deeply in debt. Never trust an Uncle who is drunk from debt.

The Old Path and Conventional Wisdom

The old-school popular advice still floats around the military: go to college, save up, find a government job, and retire. These are all good ideas, but with a hyper connected and competitive world, we need to create a bit more security for ourselves. It's not enough to make long term plans based on the way things have been. We have to actively prepare for where our economy is headed.

The Disorder Stereotypes

One of the side effects of the veterans' PTSD awareness campaigns is the aggressive stereotyping of veterans as "troubled." Because of these campaigns, many Americans just expect things to be wrong with veterans. This is something to be cognizant of as we apply to jobs, interact with people, and apply to colleges. Like it or not, "spreading awareness" about PTSD has done just that. I've run into countless people who speak to me as if I'm an injured puppy, a maniac ready to snap, or feel the need to ask probing questions to make sure I'm okay. This is an important example and illustration of iatrogenics (harm done by those who seek to help).

There is No "D" in PTSD

Let's drop the "disorder" part of PTS. I'm not particularly fond of those labeling the natural stress reactions people have to trauma as a disorder. As members of an all-volunteer military, we can decide what labels and words we use to describe our experiences and ourselves. "Disorder" doesn't have to be one of them.

The Urge to Check Out

Checking out, or strategically isolating ourselves, is tempting when facing all these challenges. But when we begin to study how we can fight these challenges on our own terms, we might be surprised at what we discover. Virtually unlimited energy awaits those willing to tackle huge challenges. There are few things more rewarding than finding ways to test our competency and skills against those whose goals are to work less, seek comfort, and acquire indulgences.

Taking Out Frustrations the Only Place We're Allowed

"Any person capable of angering you becomes your master; he can anger you only when you permit yourself to be disturbed by him." – Epictetus

This is the most dangerous threat to our families and loved ones. It's the hardest to spot, the most insidious, and the one we're not warned about. The military offers no traditional outlets to release the frustration about bureaucracy. It's well known that for anyone with a baseline of intelligence, a bureaucracy creates innumerable frustrations which can build into anger. Great organizations and traditions can descend into bureaucratic hell when they:

- Create or allow toxic cultures to evolve
- Make the threat of punishment greater than the promise of reward for speaking up about what's wrong or taking the initiative to do what's right

- Financially encourage individuals to find indulgence oriented careers
- Allow and encourage weak leaders to use indulgences to gain rank
- Force, on a daily basis, the best soldiers to watch countless transgressions while being unable to take action
- Allow speaking up about anything to possibly cause or risk severe career difficulties
- Flag a soldier for future promotion or progression by using the label "PTSD" just for talking to someone about the anger felt regarding being forced to operate under such stupidity.

Frustrations naturally build up around what we can't talk about. All too often, these frustrations and feelings spill out into the paths of least resistance. For most of us, these paths happen to be our subordinates, our families, and ourselves. It's crucial that we don't allow ourselves to punish the innocent for the frustrations a bureaucracy has piled up. It's hard to take every ounce of anger and frustration out in the gym. It's easier to vent (in every innumerable way) to our peers or our families. It's easy to drink too much, eat too much, medicate too much, or bash ourselves. We have to sublimate any frustrations we feel, and never cede control to the worst things and people in the world.

Finding the Why

"The impediment to action advances action. What stands in the way becomes the way."
-Marcus Aurelius

It might be temporarily disheartening to consider some of the things we're facing as veterans. But there is power in recognizing the current state of affairs globally, nationally, and locally. In the future, when we need that extra bit of inspiration or motivation, the challenges we just covered should be used as fuel. Viewing the world as it really is can be painful, but it's with that perspective that we can formulate the actions necessary to fix it. Then we can begin voting with our actions and taking the small steps to protect ourselves and our families, and begin rebuilding a better future.

Strategy and Action

"We have to create culture...reclaim your mind and get it out of the hands of the cultural engineers who want to turn you into a half-baked moron consuming all this trash that's being manufactured..." –Terence McKenna

"Study strategy over the years and achieve the spirit of the warrior. Today is victory over yourself of yesterday; tomorrow is your victory over lesser men."
–Miyamoto Musashi

Grit and Character Recognition Exercise

Grab a sheet of paper for this exercise or write in the book. This is an exercise to build our creativity and, admittedly, fuel the healthy portions of our ego.

The first part is to reflect on the three biggest tests and challenges we faced and overcame while deployed. What fears did we conquer, what uncertainty did we operate under, what did we witness and survive?

1.
2.
3.

Now consider what percentage of the American population has faced anything close to what you listed above? It's important that we discover what a truly elite class of Americans we're already a part of.

This is a simple exercise to illustrate that we already possess the prerequisites necessary for success. What we might consider normal is a far greater learning experience about good and evil than anything the average American has had to face. This exercise isn't to point out differences callously or casually. It's for us to remember that we often have faced far more adversity then we give ourselves credit for.

What Prompted Us to Originally Join?

This reflection is about taking the desire for ethics and virtue which led us to join the military and transfer it into civilian life. What were the ideals and values which caused us to join the military? Write down a few brief words or feelings which prompted you to join.

1.
2.
3.

So now we've identified the specific prerequisites for great accomplishment, our three examples of adversity. We have also reflected on the three things which prompted us to join the military.

Those prerequisites and ethics have never been in such high demand in technology, business, and entrepreneurship. The average veteran is already in an elite class of Americans predisposed to be entrepreneurial, willing to take risks, able to create value from nothing, and possess the ability to shift resources from low yield to high yield areas.

3

The Old Way

"I suspect the IQ, SAT, and school grades are tests designed by nerds so they can get high scores in order to call each other intelligent.…Smart and wise people who score low on IQ tests, or patently intellectually defective ones [who]…score high on them, are testing the test and not the reverse." –Nassim Taleb

"IQ accounts for what portion of our career success? The answer: between 4 and 10 percent." –Daniel Pink

The Old Credentials

The Holy Trinity of education fundamentalists in the old economy revolved strongly around IQ, SAT, and GPA.

IQ Is the Smallest Part

Researcher Daniel Goleman and the Hay Group have shown that IQ accounts for only 4 to 10 percent of our success. The mindset and conclusions to be drawn from this are that after a baseline of intelligence, it's on us to acquire what we need.

The SAT That Never Mattered

Sadly, the SAT and every other test like it measures only how well you can do on the SAT. Yale University professor Robert Sternberg has been at work for years building an alternative, or as some call it, "supplement" to the test. Sternberg's test includes:

- Writing funny captions for five blank *New Yorker* cartoons
- Narrating a story
- Addressing life challenges (ex. how would you handle arriving at a place where you didn't know anyone
- How would you try to convince friends to help you with something?)[6]

Sternberg's fledgling test has already been shown to be twice as successful as the SAT in predicting how well students perform in college. In the next few years, even more data will come out as the success disparity grows between those who focus on what matters, and those who focus on old holy grails of education: memorization and standardization.

As with IQ, after a baseline of competency, awareness, and attention to detail, it's on the individual to learn what they need. If we focus on tallying and quantifying what the predictors are of our future success, we'll stay mired in inaction, or convince ourselves out of great accomplishments. But if we become definite that we can improve and grow, we'll naturally develop the heuristics, emotional intelligence, and creativity required to succeed in life and business.

The Companies that Stopped Asking about Grades and GPAs

Google, Khan Academy, and a slew of other Silicon Valley technology companies are no longer asking applicants about grades,

GPAs, and where they went to college. At first, these institutions used the old metrics listed above to find employees out of pure convenience. Now, companies like Google have found methods which work better to find great employees, and they're embracing them wholeheartedly. Laszlo Bock is the man in charge of hiring at Google, and this is what he had to say about who they want to hire:[7]

"For every job, though, the No. 1 thing we look for is general cognitive ability, and it's not I.Q. It's learning ability. It's the ability to process on the fly. It's the ability to pull together disparate bits of information."

Laszlo went on to detail the second most important thing they're looking for:

"...is leadership — in particular emergent leadership as opposed to traditional leadership. Traditional leadership is, were you president of the chess club? Were you vice president of sales? How quickly did you get there? We don't care. What we care about is, when faced with a problem and you're a member of a team, do you, at the appropriate time, step in and lead. And just as critically, do you step back and stop leading, do you let someone else? Because what's critical to be an effective leader in this environment is you have to be willing to relinquish power."

Google is one of many companies who are interested in applicants with character, common sense, and grit. A GPA is a fine measure of prowess for memorization and devotion to curriculum. But a grade of an A+ hardly matters if we're testing something anyone can look up the answer to in minutes. A GPA of 4.0 doesn't matter if the applicant hasn't mastered technology, friendship, or the ability to learn and lead.

The businesses which thrive in the future will continually move away from old metrics of competency. They're now scouring for individuals with high emotional and creativity quotients. These are the skills at lowest risk of becoming commodities in the future. These are also the traits which make someone coachable, speed their learning, and generally contribute positively to company culture.

Silicon Valley technology companies are increasingly creating a philosophy on life and business which is spilling out into the rest of our country. Consider some of these practices that hail from Silicon Valley and are present in successful startups:

- Encouraging work done in flow states (work in the zone)
- Flattening hierarchal structures (think distributed teams such as Special Forces)
- Preferring leadership over management (leadership is management with less coercion)
- Pursuing results over fundamentalism (they're not biased about how things get done)
- Valuing emotional mastery (culture is everything)

These companies are building their own mechanisms to replace the outdated, "where'd you go to school?" and, "what were your grades?" These old questions don't really help us vet applicants unless we're applying for entry-level positions which partake in standardized tasks. When we are asked these questions when applying for a job, it means we're applying for a job rooted in old paradigms.

Silicon Valley has championed the philosophy that grades are not as important as results. I'm predicting that the startup ecosystem which inhabits most metro areas will be the next transmission mechanism for this common sense philosophy. As massive returns and success continually arise from businesses springing from startup ecosystems, prepare for this philosophy to go mainstream.

As results, skills, and interpersonal emotional intelligence and empathy become more quantifiable, testable, and preferable, expect more formalized tests for them to appear. Grades are largely a test of how much students can suspend reason, demonstrate subservience, or pretend that Soviet-style top down forced distribution and regurgitation matter. With skyrocketing tuitions, huge student debts, and difficulty measuring the return on investment, the entire system is in trouble.

There will be far less need in the workplace for those who are manically competitive in contests to memorize random facts. There will be an increasing need for workers who can repeatedly identify where and when it makes sense to compete, and where it does not. This is the essence of entrepreneurship: can you help create things which are valuable and get them to people who need them?

Consider this: an entry-level job at the federal government might require a GPA of 3.75 or above. That job probably receives hundreds of applicants a day. Those hundreds of people are competing so feverishly for a salary, armed with incentives to use outdated technology and do enough to get by. Now consider the job at a local venture backed startup where learning skills which produce results is vital to the company's survival. The upside of this job through salary, equity, and rate of learning is virtually uncapped. Even if the company fails, the worker walks away with knowledge of how business works, and skills which another venture backed startup would love to pay for.

This is exciting news for veterans who want to enter or align themselves with industries which create technology. Old-school credentials of college and GPA are slowly fading. We're getting back to the basics and back to the old interview question of, "how can you be helpful to me right now?"

How it's Always Been Done: HRSE

The best and quickest way to break into new fields, start a business, learn future-proof skills, or generally become a *monopoly of one* is through HRSE:

- Health
- Relationships
- Skills
- Entrepreneurship

This practice is foundational in that our every interaction is an output of our health. We can't accomplish anything without first being healthy ourselves. This is what will help us win the long-term game.

So let's look at some of the new prerequisites, credentials and paths which matter in the new economy. We'll also see how these fit in with the HRSE framework.

Strategy, and Action

"If you know the Way broadly you will see it in everything."
–Miyamoto Musashi

"Point of view is worth 80 IQ points" –Alan Kay

A New Way

We've seen that the old credentials which used to carry a lot of weight are eroding: degrees, family name, school name, GPA's, SAT scores. Now, outside of the top 10% of college names which still carry a premium, degrees don't matter nearly as much. These credentials will certainly hold some value as economic insurance policies. But their value is rapidly decreasing. Let's quickly review the new credentials, skills, and intelligences of the modern era:

Emotional mastery: Can we add value to teams just from being present?

Marketable skills: We know we have marketable skills when we receive job offers, or can create or help build value where none previously existed. Marketable skills include things which the market values enough to compensate us for with money. The marketable skills of the future fall into four broad categories which fit the acronym BEDS, or: Business, Engineering, Design, and Scientists. We'll cover examples of future-proof skills among those four categories in the next chapter. If we have marketable skills in any of these categories, we can go out and make money very easily. While some people are obsessed with GPA, credentials, and letters after their names, marketable skills are a form of intelligence which it makes sense to compete for.

Creativity: As everything becomes increasingly commoditized, creativity is the last great bastion of human skill to perfect and master. Can we consistently generate, sell, and implement good ideas?

Lateral thinking ability (also blends in with creativity): How can we blend ideas and concepts to create solutions to problems? Can we make everything around us a bit better by the way we blend ideas and resources? Do people know us as people who find solutions or identify problems?

Autodidact: Once we are self-taught in one skill or field of study, we can repeat the process again and again. The object is to acquire marketable skills in such a way that we become a monopoly of one. This is the hallmark of the entrepreneur.

Network: Who do we know who likes us, who we've helped, and can vouch for our abilities?

Do we have a body of work which can be cited wherever we go? This might be an online portfolio, or an app in the App Store we created. Do we have an online presence which can work and source opportunity for us while we don't work? This might be a LinkedIn profile, our own website, an article we wrote on Medium.com, or content we've created.

4

An Opportunity Unlike Any Other

"Constantly be updating your positions on as many things as possible. And if you don't contradict yourself on a regular basis, then you're not thinking." –Malcolm Gladwell

"During the past three decades, startups in the United States have created nearly 40 million American jobs, all the net job creation in the country over that period." –Steve Case

The Builders

We've covered the fact that the real innovation in our country has sprung from Silicon Valley. But no matter what the location, real innovation happens anytime humans do something uniquely human-create technology.

Technology is defined as the ability to do more with less.

Technology includes everything from semiconductors, computers, stoves, chairs, heuristics, mathematics and medicines, to letters written to a spouse or the particular culture at a business startup. In order to create any of these new things which "do more with less", we must have a mindset that is adaptive; focused on progress, growth, and compassion.

Most people overlook how technology is created. It's easy to get caught up in studying the end result of the latest multi-billion dollar technology acquisition. By becoming caught up in the end result, many people miss chances to study the alchemy and process which created the end result. So how does new technology actually get created? Is it easy?

If creating technology and innovation were easy, massive companies would not have tens of billions of dollars parked on their balance sheets. Big companies try to innovate, but it's often difficult for them to do so. This is why they acquire growing companies.

Creating new technology requires an environment and certain type of cultural alchemy. New technology can only be created in environments where participants are allowed to have new ideas and run low cost experiments to test the validity of those ideas. In order to better identify these environments, we can delineate between two main types:

- Fundamental (Bureaucracies, Government): Possessing general and vague skepticism towards "new" ideas, where sophistication is valued for sophistication's sake. Credentials and letters after names are highly valued here. Think political correctness and suits. The tribe will ostracize you for being wrong, and for taking any risk (hence incentivizing inaction).

- Anti-Fundamental (Silicon Valley, Startup Ecosystems): There is cultural alchemy here, think jeans and t-shirts. Substance over flair, the real question is how can you help right now, not what academic institution you attended. The tribe will not ostracize you for being wrong, as long as you're attempting to create.

It takes a certain type of anti-fundamentalism and freedom to create a culture in which technology can be birthed.

This book is about moving towards a much more startup minded or entrepreneurial nature. This type of pursuit requires us to have strong opinions and ambition, but it requires we let them go easily when we're proven wrong by evidence. This is the type of personal philosophy which increases the chances we'll create new technology or do great works. As Emerson said,

"A foolish consistency is the hobgoblin of little minds, adored by little statesmen and philosophers and divines. With consistency a great soul has simply nothing to do. He may well concern himself with his shadow on the wall. Speak what you think now in hard words, and to-morrow speak what to-morrow thinks in hard words again, though it contradict every thing you said to-day. –'Ah, so you shall be sure to be misunderstood.' –Is it so bad, then, to be misunderstood? Pythagoras was misunderstood, and Socrates, and Jesus, and Luther, and Copernicus, and Galileo, and Newton, and every pure and wise spirit that ever took flesh. To be great is to be misunderstood."

We must wholeheartedly trust ourselves, but be willing to find out tomorrow that we're wrong. When we find evidence which displaces our beliefs, we then incorporate that information into our heuristics, culture, or whatever technology we're creating. From that point, we set out again, and each time we get new information, we push it into an objective light. This is the type of philosophy which precedes new technology.

The fundamentalist, on the other hand, has strongly-held convictions and beliefs which cannot be questioned. The fundamentalist encounters information which contradicts his current positions and responds with defensiveness, anxiety, anger, befuddlement, confusion, or aggression. The fundamentalist chooses to see himself and his work as concretely unified. Fundamentalists prefer show to substance. They prefer bright shiny objects or cultural indulgences to technology with a high utility. They seek more for the sake of hoarding, instead of strategically acquiring goods or capital in order to use it and create wealth for others. Once anyone's thought process is mired in fundamentalism, it's incredibly difficult to shake it loose.

Sadly, fundamentalists never allow themselves to grow wise. They mistakenly associate being wrong or feelings of ignorance as precursors of fixed intelligence. When they hit this sweet spot, which indicates potential to learn, their ego shields them from incorporating new information into their existing beliefs.

Every successful and satisfied entrepreneur is someone who grows to despise fundamentalism. These entrepreneurs, or founders, are staunch anti-fundamentalists. Let's consider some of the first founders (in every sense of the word), our founding fathers.

Our country was created when a group of cultural alchemists, tinkerers, and anti-fundamentalists got together to create new technology. The type of government these risk takers created was a piece of new technology: a democratic republic. Our founding fathers could only birth such a miracle by associating with other like-minded creators. The founder DNA, or allele for you biologists, runs in the blood of every American.

We live in a world where every piece of information and knowledge is available at low cost to us online. Businesses have access to a global pool of people who are smart, wise, hungry, able, and ambitious —and who all look up to America.

Sometimes the difficult part of the equation is getting to a place where we're allowed to fund, tinker, create, and sell new technology. If we try to create technology among a bunch of fundamentalists, they'll point out everything we're doing that is *wrong*. They will smother creations before they can be birthed. This is why it's incredibly important for us to go to places where it's culturally acceptable to create. These places are the startup ecosystems popping up all around the U.S. These are cultures where personal growth is the norm, instead of taboo. These are ecosystems where those who attempt to create and take risks are revered.

These cultures offer the opportunity of a lifetime for those who embrace personal growth. Right now, as you read this, there are people knee deep in these areas of growth seeking to make things better for humanity on a massive scale. The blessing and curse of the 21st century is that we're able to view how the work we do ties into bigger ideas. Is our work creating new technology? Or are we simply spreading old technology and ideas all around? The blessing of this phenomenon is that we're allowed to join or align ourselves with industries creating technology.

As the Internet and technology continue to proliferate, we'll have to watch these innovations happen—with or without our involvement. Now, let's quickly cover the types of industries which are creating technology.

Industries Poised to Explode

3D printing, robotics, drones, technology, same day delivery for groceries, agriculture, books/learning materials, electronics, wearables, sensors, crypto-currencies, space travel and exploration (even terra-forming Mars), engineering, design, private equity, driverless cars, energy, software and hardware development, artists (creatives skilled at their jobs, even including sales people), preventive medicine, future-proof trades such as welding and plumbing—all these fields offer huge opportunities for those who are entrepreneurial.

By entering these fields with the proper strategy, they all offer potential for uncapped upside (this simply means opportunity and wealth way beyond a traditional salary). That strategy and mindset includes recognizing that in our new economy, there are really only four types of skill sets.

The Four Categories of Skills These Industries Desperately Need

There are four types of cornerstone skills and professions in the new economy. These skills are in such high demand we can often get a job doing them after only a few months of training. These categories of skills encompass just about every type of future-proof job.

Business Developers: sales, growth, strategy, marketing, social media content, full-stack marketing, general leadership, storytellers, authors, artists, etc. (Artists who want to survive in this transitional period where new media emerges as a dominant force must absolutely become business people). The "retirement" of business in our modern age will likely be politics. For veterans interested in politics, this is the place to start.[8]

Engineers: mechanical engineers, machinists, tradesmen, developers, front-end, back-end, full-stack, and the whole makers movement.

Designers: UX/UI, HTML/CSS, Photoshop CS, social media content, wireframes, sketching, rapid prototyping, 3D printing. This also includes anyone who creates art.

Scientists: data science, traditional scientists, statisticians, computer scientists, biologists, and medicinal researchers. These professions require a mastery of manipulating, understanding, and making prescriptions based on huge sets of data. Real scientists will be anti-fundamentalists.

The term "business development" is a good overall description of what is becoming demanded of salespeople who are also technical. Business developer experts usually become the founders of enterprises. Founders can certainly come from other domains, but they most often emerge as masters from several disciplines: i.e. business and science, or design and business.

How to Get There

These four main skill sets are professions ready to welcome veterans with the right skills. As we've said before, acquiring the specific skills to gain entry to these jobs is much easier than the traditional 2 to 5 or 6-year degrees. These positions can all be accessed via short term (3 - 12 months) bootcamps, apprenticeship programs, or intensive training for the skilled trades that are in high demand such as technical manufacturing, plumbing, welding. These are all accelerated entry points into true technology fields. The demand for the value these fields are creating warrants accelerated training times, high pay, and a high degree of learning on the job. If we remember the proper definition of technology we covered earlier (the ability to do more with less), fields that work with technology includes everything from coding, to technical, manufacturing, welding, plumbing, or data science. All these fields do more with less and prove it to the prospective applicant by nature of their high salaries and straightforward training, allowing a newcomer to jump right in.

If we think attending one of these is silly because the Post 9-11 GI Bill doesn't cover it, we might want to reconsider. Not only are there alternative funding sources for these opportunities, there are many bootcamps where tuition is paid by the employer who hires us at the end.

Some of these bootcamps charge up-front, and some of these bootcamps only charge after we've been placed at a great job. Reputable bootcamps have placement rates for graduates of around 95%. The average starting salaries for graduates range from $60,000-134,000 and in many cases come with incentives such as equity, stock options, and room for explosive growth at startups. So not only are these salaries amazing, but now we've started to expose ourselves to high rates of learning and upside in the form of equity. Owning a few percentage points of a startup by joining as an early designer doesn't sound like much, until the startup is acquired for a hundred million dollars.

The companies hiring employees coming out of bootcamps are on the cutting edge of innovation. Veterans are perfectly poised to thrive amid the uncertainty, high operational tempo, and the challenges these routes offer.

When thinking about the bootcamps which provide amazing opportunities, feel free to interchange the term "bootcamp" with any program or training that offers the quickest entry point possible into high demand fields.

Reintegrate, Checkout, or Rebuild?

Amazing work is available, but only for people who have skills. There are no good old boy networks, no secret handshakes in these fields, just results. These fields are meritocracies, in businesses with increasingly flat hierarchal structures. This is a rare opportunity to find work which ties into huge, audacious goals like ensuring the future and survival of humanity.

All the previous generations of our ancestors fought to give us chances like this. Our country's founding fathers took risks so we could do the same, not so we could kick back in comfort. The opportunities we have access to today are those which our grandparents couldn't have dreamed about.

Although it might be a struggle to get into these fields at first, it will be even more of a struggle if we spend the next four to six years learning outdated material in the form of a Bachelors, Masters, or PhD.

While the U.S. GDP stands at around $17 trillion, it's important to remember that the types of companies which create technology access the world market. Becoming competent and aligning ourselves with technology helps us to access a global $70-trillion-dollar economy. This is an opportunity to rebuild America and lift up ourselves and our families in the process.

One of the hard truths about new technology is the fact that so much of what we do is easily accessible and recordable. Our actions go on record. How and what we use our actions to vote for across the span of our lives is now being recorded.

Our children will see everything we do or don't do. They'll see the principles we stood up for, what we gave our lives to—or perhaps they will wonder why we never took advantage of such tremendous opportunity.

Haven't Veterans Done Enough Already?

Isn't asking veterans to do and risk more presumptuous? Maybe —and if it comes across this way, I apologize—but I selfishly want the future leaders of our country to come from the ranks of our veterans. America is in desperate need of leaders.

I want my family to live in an America where those who have taken personal risks to create technology are the ones who hold positions of power. I don't want to look for "leadership" to those who have never taken personal risks. I want to see veterans on the news changing the world. I want to buy goods and services from veteran-owned businesses. I'd be willing to bet we all want something similar.

There is certainly still opportunity in traditional paths such as staying in the military, or working in government or law enforcement. This isn't a call to completely avoid these professions. There is no reason we can't go into those professions, then enter politics or start a business. But this is a call to become very aware that these are the skills which will dominate the future of our economy. This is a call to decide for yourself what your goals are and where your interests and talents lie.

A Virtuous Doxastic Commitment

In philosophy, there is a type of belief called a doxastic commitment. A doxastic commitment is a class of beliefs so strong it naturally causes the individual holding it to take actions based on their belief. As veterans in an all-volunteer military, we're familiar with having beliefs this strong. Doxastic commitments are what led us to joining the military.

Now we have an opportunity to do so again, and join those rebuilding America. In the process, we get to take advantage of massive upside, protect our downside, and be on the side of American history which owns the process of restoring our republic.

We are of a class of people who makes virtuous doxastic commitments. We back up our beliefs via action.

Strategy and Action

"A goal is not always meant to be reached. It often serves simply as something to aim at." –Bruce Lee

"I knew that when I was 80 I was…not going to regret trying to participate in this thing called the Internet that I thought was going to be a really big deal. I knew that if I failed I wouldn't regret that, but I knew the one thing I might regret is not ever having tried. I knew that that would haunt me every day…" –Jeff Bezos

Four Central Professions and Minimizing Regrets

Following are some brief thoughts and questions to strategize on how best to think about the opportunity in front of us.

We see that the developing world is still looking to us for innovation. Many people in these countries are speaking English. The industries we covered, which are growing now, will only explode in growth as they begin to serve a developing world which is coming online and reaching the affluence levels where they can afford services from these American companies.

Where do we see our natural talents and skills in regard to the four central professions of the future? Maybe we want to stay at our current job because of the retirement or our kids, but how could we learn and layer some of the skills below onto our existing skills sets? Where do our natural abilities and interests already lie?

Of the four main skills below, where do we see ourselves already being a good fit? What do we feel pre-qualified to learn or layer on our existing skills?

Business Developers:
Engineers:
Designers:
Scientists:

Every company which aims to survive is increasingly becoming oriented around creating technology. The fields above can provide us with meaningful career certainty, options, and the ability to start our own technology company after we've worked in several successful (or unsuccessful) ones. There's a lot to be learned from going through what *doesn't* work.

How Will Posterity Remember Us?

Write out how you think they'll remember us: the generation of veterans who rebuilt the country. How, in detail, do we want our children to reference what we've created after we're gone? How will they view veterans, and how will they remember us? Will they remember us as the generation of veterans who helped rebuild our country? Will they remember us as those who chose reinvention over reintegration? Will we become the leaders and entrepreneurs who completely dominate these industries (even politics)? Will we be the ones who shape and preserve these institutions?

Part 2: Health

5

Functions, Culture, and Growth

"This is the very perfection of a man, to find out his own imperfection." –Saint Augustine

"The best people, they're afraid, they question themselves. Many, if you corner them, will admit that they wonder if they're good enough. But what separates them from the rest is that they jump off the cliff anyway. Sprout wings on the way down." –Kamal Ravikant

On Functions and Agency

There are tens of thousands of companies and individuals fighting to rebuild America. The developing world is coming online, and they're purchasing more and more goods from the West. The standard of living in the United States has never been higher. We have access to technology and infrastructure which businessmen living a decade ago would have killed for.

But in order to fully capitalize on the possibilities in front of us, in order to find, join, or personally create the type of work and value which will rebuild America, we first have to adopt a mindset of complete personal agency. Personal agency means we accept full responsibility for ourselves; everything we do, consume, act on, or fail to act on. As we've no doubt noticed, there are many people who go through life terrified of accepting full personal agency for themselves.

Accepting that we control many things in our lives spotlights the absurdity of excuses. It makes our past selves look weak, and this is okay. We'll need to constantly create and ship updated versions of ourselves if we're going to rebuild America. Personal agency is not to be confused with irrationality or hubris. There are, and always will be, many things outside our control. This collection of things and random events can be summed up as fate. Fate will always rear its head. We've all heard Murphy's Law:

"Anything that can go wrong will go wrong."

The person brave enough to accept full personal agency accepts Murphy's Law, but also proudly lifts a middle finger to fate. While they flip off fate, they ruthlessly prepare, evolve, learn, and seek out secrets to protect themselves from whatever fate may bring. They plan, despite knowing full well that their plans could disintegrate with first contact with an enemy—or the marketplace.

Fully accepting and leveraging personal agency is a practice which takes years to master, but once we do, we wield a power that looks mystical to those who bow their head to fate.

Either we go all in on personal agency or we don't. Either we spend our life and creativity finding excuses, people, and institutions to blame or we'll figure how to re-create them in a better image. Either we're waiting for a handout, the lotto, stars to align, a fully functioning business to fall in our laps or we're busy creating and living our lives.

I'll admit I had a natural aversion to this philosophy for a long time. The truth is it takes a certain level of courage and faith to accept our own agency. But once we accept that we have control over our lives, we become free. We're free to pursue and achieve our dreams. We're free to create a legacy for ourselves and our families. We're blessed with the joy which comes from achieving things we formerly thought impossible.

Full personal agency begins through a mastery of our health. By optimizing our health, we are in the best position to mitigate what fate throws at us.

The level of our own health sets the stage for every future thought, conversation, relationship, skill, and achievement. We're figuratively putting the oxygen mask on our face and breathing in deeply before we attempt to save anyone else.

When we take full personal agency in our lives, we can't help but create the type of lives other people view as "lucky." Lucky is just an explanation for day in and day out preparation. Luck is the word the entrepreneur uses when he knows most people aren't interested in listening to his 12 hour back-story describing his success. Luck is the word the veteran who landed a spot on the founding team of a venture-backed startup uses to describe how he got the job. Full personal agency founded on personal health naturally creates "luck."

"The man on top of the mountain didn't fall there."
–Vince Lombardi

The Function of You

The most valuable mindset which accompanies personal agency, is recognizing that we are each a function.

In math, a function is a relation between a set of inputs and a set of outputs. When we view what we're trying to achieve as simply a function, the process of getting there becomes demystified. Everything we allow into our lives, our bodies, and our relationships is an input. It is all directly responsible for the resulting outputs. If we want to change our outputs, or what we create, we only have to properly value and filter the inputs.

The first part of creating a great function is deciphering and planning for the exact output we want. This output can change many times, but it's important to have a direction and aim. We can work towards something temporarily and decide it's not for us, or keep pushing forward to find out if it is. The point is we need an output which we are trying to create. That output could be a new skill, healed relationship, mastery of our health, six-pack abs, starting a business—whatever we want. The idea is to select outputs large enough that we become motivated to find the inputs needed to create them. The function mindset asks us to think global and act local.

Having an output we want to create helps us get in motion. Once we are in motion it becomes increasingly easy to identify and pick out the inputs we need and adjust them accordingly.

Right now, the outputs we've created in our life are from what we've consumed. We are a function which has created what we've taken in. We are the food we eat. We are the people we hang out with. We are what we read. We become what we think about. Everything we've been able to produce, or have failed to produce, is a result of our inputs. If we don't take proper care to monitor and adjust our inputs accordingly, we will fail. So how do we know what inputs we should be taking in?

In order to get the output we want, we simply need to find people who have created that circumstance. Analyze them and find out how they created what it is we value. We can check to make sure they aren't suffering from a bunch of maladies or side effects of creating what we want. Then, we study the type of inputs they used to get there. They've likely read hundreds of books; we can find out which ones and then read them ourselves. They probably exercise; what kinds of workouts do they do? If they look healthy, what is their diet like? Who do they spend time with? Who did they start their business with?

When we know the outcome we want to create, there are only two types of inputs we interact with. Those two types of inputs are: ones which help us, and ones which don't help us create the outputs we want. If any inputs fall into a seemingly grey area, be wary of them.

The foundation of everything we do starts with preventative health. Our society places an absurdly small value on the preventative medicines of diet, exercise, and sleep, yet these are the actions which produce the greatest benefits. By focusing on health first, we boost every interaction we have with others. When every interaction with others becomes better, we can't help but continually move our lives forward. When we shield ourselves from dangerous inputs and continually make small experiments to improve our health, our entire lives can be transformed. We regain control of our brains and we boost our willpower. This practice pays dividends for the rest of our lives.

For instance, consider something small such as eliminating soda or diet soda from your diet and drinking more water. This begins to bring our bodies Ph levels to a normalized state. It stops destroying our bodies' resistance to insulin. The lack of sugar (assuming we cut it other places as well) allows the inflammation it's caused in our brains to heal. We then begin to reduce our risk for diabetes, Alzheimer's, depression, and a whole host of other maladies. We start to win back control of our brains, which then influence everything else. You might be thinking this sounds a bit extreme, but it's not. This simple act of cutting sugar from our diet can mean the difference between 60 more years of life with a healthy brain, or a slow decline into dementia. We are a function of what we consume.

When we break down our health into a function mentality, where we put in inputs and produce outputs, it becomes easy to know where we're heading. There aren't any secrets about what we're going to achieve.

There are two types of inputs: raw materials and our actions. These give us our outputs. By becoming obsessed with the raw materials we consume, we give ourselves energy for the necessary actions and willpower needed to produce outputs we want. Now, we need only to figure out the outputs we want to create, and consume and act appropriately.

Strategy and Action

"A man's worth is no greater than the worth of his ambitions." –Marcus Aurelius

"The man who does not value himself, cannot value anything or anyone." –Ayn Rand

Inputs, Outputs, and Emulation

The function mentality is simple, yet challenging to implement daily. The first step we can take towards successfully executing on the function mindset is to become certain about the outputs we want to create.

We can start by listing the outcomes we want for ourselves. These could be health, satisfaction (more than happiness), developing the ability to do the type of work we love, being promoted, owning our own business, becoming a multi-millionaire, building an amazing family, or a combination of the above. The point is to get very specific about what we are creating. When we're hyper specific about what we want to create, it becomes easy to find others who have created those things. When we find those creators, we can then discover and study their inputs and actions. After we identify these, it's only a matter of consistently executing on those inputs and actions. Our minds need specificity. Our faith in where we're headed needs to be kindled by proof that it's possible. So let's get started by listing the outcomes (outputs) of a function we want to create:

1.
2.
3.
4.
5.

Now we find who has created these outputs in their personal lives?

1.
2.
3.
4.
5.

What are some of the inputs, mindsets, and actions they cite or have pursued (sometimes behind the scenes) in order to create what they have?

1.
2.
3.
4.
5.

If this process sounds simple, it is. The thing about simple processes is—most people won't follow through with them. Aspiring to greatness requires courage, faith, and exposing ourselves to failure. The people who start this process but don't finish are those who worry too much about failure instead of worrying about what it would be like if they never tried.

The lesser portions of our ego can trick us into thinking that finding someone to emulate means we're not good enough. This type of thinking is poisonous. Don't let it stop you from finding outcomes and emulating successful people.

The type of people who view intelligence and capacities as fixed, are toxic. In the book *Mindset,* Dr. Carol Dweck discovered a simple difference between those who achieve what they want, and those who do not. The difference between the two types of people was that one group saw their abilities as *fixed.* They bought into the concept that our brains and capacities remain the same, and we can't change them. These types of people (surprise!) never changed in life. The type of people Dr. Dweck found who were high achievers had a mindset of *growth.* They saw their abilities, brains, and bodies as malleable functions they could control. They naturally achieved what they wanted.

Those with fixed mindsets never try and never learn. Those with growth-oriented mindsets sometimes have hubris and unwarranted confidence, and sometimes they under- or over-shoot, but because they get started they're guaranteed to learn. Those with a growth mindset are bodies in motion, and once bodies are in motion it's much easier to change direction and adjust course.

Caveats on Finding Those to Emulate

I understand the objections some people have with these ideas. They don't want to be exactly like the person they're emulating, they have their own talents and ideas. But finding a person to emulate isn't about becoming exactly like them. It's not being slavish or a copycat. Finding a person to emulate for the outcomes they've produced is the best way we can discover ourselves. It tests us strictly and guides us on a path which has worked before. It's through getting started that we'll discover what feels authentic to us, and what does not. It's when we mix emulation with thoughtfulness that we find originality.

Once we find a person or group of people who have produced what we want, we can then determine if they've been able to maintain the other parts of life we value. So maybe they have made a million dollars, but at the expense of their relationships. Their grown kids won't talk to them; their girlfriends have orders of protection against them. Is that what we want? Hopefully the answer is no, so we go on to the next person. Maybe we can find someone who has had more modest business success, but maintained their relationships and health, or someone who is building their wealth at a sustainable pace. We can continually sift through those we want to emulate until we're left with people who have produced what we want and haven't sold their souls in the process. Once we find these people, we can further sift through this group to identify people who came from similar backgrounds as us. For instance, if our parents got divorced when we were in our teens and we can find somebody to emulate whose parents also got divorced in their teens, that person's experiences and inputs will likely be much more contextually relevant to us.

So the process of finding the type of outputs we want to create starts like this:

1. Discover the outputs we want.
2. Make sure we have a solid "why" for each output we want to create. If our "whys" are big enough, we'll be much more inclined to figure out the "how."
3. Find several people who have produced outputs similar to what we want; these are potentially people we want to emulate. (List them.)
4. Use due diligence to try and figure out how they've been able to achieve what they have. Have they sacrificed certain areas of life along the way? Have they maintained balance? This due diligence exists to make sure we're emulating those who have produced the outputs we want in a healthy way.
5. Filter out those who have produced the outputs we want and have remained great people while doing so.
6. Filter out those people again to find those with backgrounds most

like our own. The more similar their backgrounds are to our own, the higher the probability that the inputs which mattered to them will be contextually relevant to us.

7. The list we're left with is where we can start searching for the exact inputs and actions they've used to get the outcomes we want to achieve.

We don't have to contact these people directly, but the people we're left with are our ideal "digital mentors." If these people have a presence online, it will be much easier to decipher their inputs and actions. Study their backgrounds. The more closely your background and character traits match up with those you're emulating, the easier it will be to relate.

The more similarities we can find between their early life and our own, the better. Maybe they grew up with a great family or maybe their parents were divorced. Maybe they were an only child, or they lived in a house packed with brothers and sisters. Maybe they went to college and found it vital; maybe they dropped out of school. Study all these things. Once we learn these facts, study them even further to find their interests, habits, addictions, or weaknesses. When we get that specific, we will have found someone worth emulating, and the perfect digital mentor.

It's in this pursuit that we have a high chance of stumbling over opportunities in our own lives which are unique to our circumstances.

No matter what stage of life we're in, there are people who have the health, wealth, and wisdom we want. This is the first time in history where many of these people are sharing their experiences and knowledge online. Many people want in-person mentors, and in-person mentors can definitely be helpful. But if we can expand our minds from demanding in-person mentors to learning to emulate the best via digital mentors, we expand our horizons. Now, we are able to receive virtual/digital mentorship from the best outcome producers in the world.

Now that we have an idea about the outcomes we want to create, we can dive into more of the inputs needed to get there. Some of the most important inputs are the raw materials we consume. These raw materials give us the energy and fuel to take the actions necessary to create the outcomes we've listed above.

6

A Foundation

"Tell me what you eat, I'll tell you who you are."
–Anthelme Brillat-Savarin

"Researchers have known for some time now that the cornerstone of all degenerative conditions, including brain disorders, is inflammation. But what they didn't have documented until now are the instigators of that inflammation—the first missteps that prompt this deadly reaction. And what they are finding is that gluten, and a high-carbohydrate diet for that matter, are among the most prominent stimulators of inflammatory pathways that reach the brain." –Dr. David Perlmutter

Lunch with Kevin Part 1

"It's f*#&ing really nasty stuff," said Kevin.

I laughed in Kevin's face. He widened his eyes and made a finger-down-the-throat motion. His neck muscles bulged. I try to stay in good shape, but I couldn't help but feel like an amateur every time I saw Kevin. Just hanging around him made me remind myself I needed to get to the gym three days a week instead of the two half-assed sessions I'd been putting in.

"Like, know for real—the reasons are f*#&ing nasty. Learning that shit once will really f*#&ing change your game."

We were eating lunch at an upscale restaurant in Washington, D.C. and Kevin's voice was uncomfortably loud. The expletives he causally dropped made me glad the restaurant was largely empty. But on the other hand, I didn't think anyone was going to say anything to Kevin. Kevin looked like a cross between an MMA fighter and a running back. He was jacked, to put it mildly. He had his PhD in exercise science, but had become obsessed with nutrition and health several years ago. He was the only trainer I'd ever met who didn't have to sell his clients on the idea; he knew what he was talking about. Kevin started his practice around 10 years ago, building it from the ground up. Now he had a podcast and was getting ready to sign a lease on a gym which would double as his office.

We rarely got to catch up, and as we sat waiting for our food, I took full advantage of the massive amount of knowledge he had acquired in the health space. I took another sip of my lemon water and peppered him with questions.

"I'm almost sold on the diet stuff you're talking about," I said. "I mean, I'm practicing a lot of it now."

He leaned over as if he was broaching some taboo subject. "How long have you been off that shit?" he asked.

I laughed at his allusion to drugs, but I knew what he was referring to: sugar and gluten.

"Well, I guess it's been a bit over three weeks. Occasionally I'll lapse," I said.

"That's decent, but you're f*#&ing shortchanging yourself. You're not going to get results till you're a full month COMPLETELY clean." He leaned back and folded his tattooed arms like an angry parent.

"I know.. I understand the basics behind it, but if you had to sell me on the holy grail of health—what you sell to clients—what would you tell me in a couple minutes. The risks, the strategy, the why. You know what I'm saying?" I asked.

"Yea, yea, yea, I get where you're going with this but understand you're at a point where you want to believe. Plenty o' people are not at a place where they want to believe. They just want pills and supplements to fix the symptoms. The real answer is the f*#&ing truth, but it's tough. You better not be f*#&ing counting yet!"

Kevin pointed to my watch like he was going to rip it off my hand.

I motioned to the watch. "I'm not counting…YET. But, yeah, I mean let's hear it—why should I be gluten and sugar free?" I tapped the table and pointed to him, as if to say time starts now.

"Gluten is Latin for glue. In grade school the kids that ate too much glue got messed up. It's no different now. The human intestines and brain are not meant for large or steady amounts of sugar, grain or gluten. They all cause inflammation and are starting to be linked to just about every modern ailment imaginable. If you don't want dementia and a leaky gut, you don't want to eat them. Need one place for all the proof? Read Dr. David Perlmutter's *Grain Brain.* That will get you completely clean." Kevin put his hands up as if to say that's all I've got.

Needless to say, I checked out *Grain Brain.* If there are three inputs we can remove from our diet to completely revolutionize our health, they are sugar, gluten, and grains.

It's that easy and it's also that hard.

What's the problem with the old Western diet? Look at the results: America is facing an epidemic of obesity, diabetes, and young people crippled with conditions caused largely by processed foods and a flawed food pyramid pushed on us from birth.

Even worse, many of us are not tapping into our full potential, learning at a quick rate, or accessing our full stores of willpower. When we consume less than optimal raw materials, we make it unnecessarily difficult to get what we want. Consuming raw materials which pollute our bodies and destroy our brains is like driving a brand new car everywhere with the emergency brake on.

The food pyramid was, at best, a clever trick by the corn growers' lobby, government, and certain doctors to get people eating more grains. In 1992 the U.S. government recommended a low fat, high carbohydrate diet to all Americans. The American Heart association followed suit in 1994. The food pyramid posters and "low-fat" products went out en masse everywhere. In 1990 there were around 6.4 million diabetics in the United States. In 2009 the number of diabetics exploded to around 21 million. Now, a person in the United States dies from diabetes around every six seconds. These staggering numbers don't even cover all the other physical and neurological ailments which arrive before and after diabetes.

The goal is to build a diet more in line with that of our ancestors. We want to start with a diet that mimics a Paleo or "slow carb diet." These are diets more in line with what humans were used to for tens of thousands of years. History is our ally in creating the best diet from an evolutionary standpoint.

What to eat:

- Fish that hasn't been farm-raised
- Grass fed meats
- Healthy fats, preferably Omega 3's
- Vegetables: when in doubt, go dark green, and go organic
- Non-Sugary Fruits: avocados, olives, eggplant, lemons, limes
- When we eat starch, choose the occasional sweet potato, lentils, or other legumes

What not to eat:

The bottom of the food pyramid:

- Whole Wheat
- Carbohydrates
- Sugar
- Gluten

What to drink:

- Liquids which have been tested for thousands of years:
- Water
- Coffee
- Tea
- Wine
- Lemon water: a shot of apple cider vinegar with lemon juice, and a teaspoon of honey in 8-16 ounces of water. As we detox ourselves from sugar, we can slowly remove the honey.

What not to drink:

- Fat-free milk
- Soda
- Fruit juices
- Sports drinks

- Energy drinks

What is unnatural and new usually hurts us, and what has been tested by time and evolution is usually much healthier. For instance, consider liquids such as water and coffee. They have been around forever; we know what we're getting with each one. Now consider something like orange juice, which is just one more way for us to take in a huge amount of sugar under the guise of healthy eating.

It makes sense to begin by mastering these inputs because they affect everything: our thoughts, our brain and heart health, or lack thereof.

The Paleo craze has rocketed through the nation several times already. In the last few years, there has always been some kind of Paleo diet book on the *New York Times* Bestseller list. There are thousands of resources online to begin creating and testing a diet more Paleo in nature. We don't want to harp too much on the details here, but just give an overview of the best fuel for our bodies.

Starting anything new requires a huge amount of willpower. Instead of a complete diet overhaul which takes monumental effort to create, sometimes it makes sense to get a small win first. A simple start towards a complete diet overhaul could be consuming as little sugar as possible for 30-90 days. The effects of this small win will likely be enough to spur on the next undertaking. Many people set themselves up for endless frustration by trying to do too much at once. A small victory is better than getting overwhelmed and quitting completely. A 90 day victory of swapping out sugar for healthy fats is still a victory.

So if a diet low in sugar and high in healthy fat is so effective, why don't more people get started? One reason is that eating sugar and grains give us an immediate opiate like reward, which make them difficult to kick.[9] One of the most powerful things we can do to reduce the sugar, grain, and gluten in our diets is to study the output they produce.

The very first outputs grain and gluten produce are depression and headaches. Our brains grow inflamed and all our energy is spent fighting these poisons. Then we become sick. Then diabetes follows. If we keep going, we'll be on our way to dementia. The full outcomes and symptoms which sugar, grain, and gluten have been linked to are as follows:

- ADHD
- Anxiety
- Chronic stress
- Chronic headaches
- Migraine headaches
- Depression
- Diabetes
- Epilepsy
- Focus and concentration problems
- Inflammatory conditions and diseases including arthritis
- Insomnia
- Intestinal problems, including celiac disease and gluten sensitivity, and irritable bowel[10]

We can never expect a calm mind, decreased anxiety, less anger, fewer ups and downs, or more pronounced joys unless we control the inputs which fuel our bodies and brains. Carbohydrates, gluten, and sugar also deplete our brains of neurotransmitters vital to the regulation of mood. Consider this,

"when your blood sugar increases, there's an immediate depletion of the neurotransmitters serotonin, epinephrine, norepinephrine, GABA, and dopamine. At the same time, B-complex vitamins, which are needed to make those neurotransmitters (and a few hundred other things), get used up. Magnesium levels also diminish, and this handicaps both your nervous system and liver."[11]

For more science behind how gluten and grain affect our overall health and potential, read *Grain Brain* by Dr. David Perlmutter. The book profiles dozens of studies on the degenerative effects of carbohydrates, gluten, and sugar.

Now in the Strategy and Action section, we'll discover several 30-90 day experiments we can run to discover what getting clean from poison and opiate addiction feels like.

Strategy and Action

"Study after study shows that high cholesterol reduces your risk for brain disease and increases longevity. By the same token, high levels of dietary fat (the good kind, no trans fats here) have been proven to be key to health and peak brain function." –Dr. David Perlmutter

"Everyone thinks of changing the world, but no one thinks of changing himself." –Leo Tolstoy

The Stakes and Incentives for Taking Action

There are two levels to the sugar and gluten detox. The first thing we want to do is make the experiment incredibly simple and easy to win. We want to set up a test so we can see the results quickly. The challenging part is getting clean from sugar and gluten, and it "may take several months to see dramatic improvement."[12]

I highly recommend placing a year-long calendar on the wall. Use a checkmark to signal each daily victory. First aim for a day and signal it with a check mark. Then go for a week, and then 30, 60, and 90 days. Convincing and selling ourselves through a series of small victories is the only way we'll follow through with this. Read *Grain Brain* and talk to others who have been sugar, gluten, and grain free for several months.

An app like Lift.do is perfect for embarking on this (or any other challenge) in a community. Lift.do offers users instant access to communities of people who are pursuing similar goals. Alcoholics Anonymous works because of the community and rituals it provides.

Lift.do offers the same opportunity to get clean from opiates, er… um… sugar and gluten in a group setting.

This is the place where we can find others we can relate with. These communities of others going without sugar and gluten will all be facing the same temptations, relapses, and neurotic challenges. The Lift app helps tracks our progress in a public way. By using our name and interacting in the community, we instantly set up stakes and incentives to get clean from sugar and gluten and showcase our progress publicly.

Incentives are defined as selling ourselves on why we're doing what we're doing. We might use cultural quirks, money, telling our kids or spouse we're going to do something, etc.

Stakes can be thought of in this way: we talk to everybody in the community at Lift.do and get to know them. We then become embarrassed if we slip up. More extreme stakes are sometimes necessary. Example: write a check to a nonprofit or political group that does work we don't agree with. Then, put it in an addressed envelope, and give it to a trustworthy third party. If we make our goal, the check doesn't get sent in. If we do not, consider the money *donated*.

None of these pursuits are easy. But the upside these experiments provide—the ability to learn more quickly, have healthier children, avoiding diabetes, inflammation, leaky gut, Alzheimer's and every other ailment—makes it too important to not at least test.

A few final thoughts for the no sugar, no gluten experiments:

The empty and harmful calories from the sugar and gluten we've eliminated will need to be replaced. These calories might come from healthy fats, vegetables, grass fed beef, or fresh meat. Get crazy with it and make jerky, fry fish in gluten free batter, make sweet potato fries, eat natural bacon and sausages. Eat cauliflower, olives, avocados, etc. Cook things with coconut oil. Don't be afraid to let our fat, protein, and vegetable consumption go through the roof. If we consume enough calories from healthy sources, we'll fight the incredible cravings that arise from detoxifying from sugar and gluten. Plus, we lower our risks of relapse. If we're embarking on the challenge in a community like Lift.do, we'll be able to share pictures of the food we're eating instead of garbage.

By the way, relapse is okay and to be expected. Sugar and gluten produce a neurological output in our brains similar to that of opiates, so there is no shame in relapsing. The important thing is to not be discouraged. One slip up just means you have one less relapse between addiction and freedom.

Tips to make it through: try different coffees; buy really, really delicious coffee or get a coffee subscription to bring in novelty and take our minds off the cravings. Have plenty of healthy fatty snacks on hand, and drink WATER. In general, just be curious about what it would be like to be three months clean. Oftentimes curiosity is a powerful enough motivator to help us summon the willpower to make it through challenges.

If we've gone for years polluting our bodies with these substances, getting clean for 30 days is a serious accomplishment. The results of this challenge after 30 days, 60 days or 90 days will be enough of an argument to keep people clean for the rest of their lives.

The results and benefits on mood, mental acuity, skill acquisition, speaking ability, cognition, sex drive, ambition, and better sleep make the completion of these experiments market themselves.

7

Endogenous Dosing: Roots vs. Symptoms

Aerobic exercise in particular not only turns on genes linked to longevity, but also targets the BDNF gene, the brain's "growth hormone." –Dr. David Perlmutter

Action is what unites every great success. Action is what produces results. Knowledge is only potential power until it comes into the hands of someone who knows how to get himself to take effective action. In fact, the literal definition of the word "power" is "the ability to act." –Tony Robbins

Lunch with Kevin Part 2

We just finished our lunch, and Kevin's penchant for swearing hadn't gotten us kicked out of the restaurant–yet.

"So what's the biggest mindset difference between those who get completely healthy and in amazing shape like you, and everyone who doesn't?" I asked.

"Simple. It's a symptoms vs. prevention mindset," said Kevin.

"Go on…" I said.

"Okay, so nobody wants to hear the f*#&ing truth. Like very, very few people I coach or work with listen at first. Eventually, though, they buy into things. It's the types that trust enough to do, not wuss out and, 'try.' People who basically value themselves enough," said Kevin.

"So something clicks for them where they go for it. But what about symptoms versus prevention, what do you mean by that?"

"Honestly, man, there is a lot of stuff we can do exercise-wise, but the biggest f*#&ing things are ALL prevention," said Kevin.

"We've covered diet. You're talking about sleep and exercise, right?" I said.

"Bingo, man. It's not fancy enough for most people. They want some super complicated answer or pill. So they either always look outside, or they realize the best drugs are waiting for them after getting their sleep, diet, and working out straight."

Kevin's brow looked like he was ready to yell this from the rooftops.

"So the people that value prevention and themselves enough are just the ones who get enough sleep?" I asked.

"The clients that listen and get on board: they sleep enough, they squat, they dead lift, do something for core, and they sprint or do a variation of it. It's that f*#&ing simple. That's all the f*#&ing prevention you need," said Kevin.

"Now make sure you get back to dead lifts and squats, okay?" he said in a mock-parental voice.

"Ha-ha, yeah yeah, I will," I said.

Endogenous Dosing

We've seen that sugar and gluten produce an opiate like cycle of addiction in our brains and bodies. If we embrace a function mentality of ourselves, the next question that naturally follows is, "how do we create healthy addictions and cycles in our bodies and brains?"

The answer to that is by endogenously dosing the right drugs at the right times that our body *can* naturally produce. When we add drugs from outside our bodies, we get hosts of unintended side effects and consequences (this is referred to as exogenously dosing). In comparison, when we allow our bodies to endogenously dose the right amount of drugs at the right time, through the right amount of sleep, diet, and exercise, we give ourselves exposure to tremendous upside. This upside includes the ability to literally create better genetics that we can pass on to our kids (epigenetics) and allows us to turn on growth factors in our brain that can do everything from speeding up learning to facilitating the growth of new neural stem cells. This chapter is about finding the right actions to mix with our inputs so we can endogenously dose what our body needs for the long term.

The first type of inputs we talked about are some of the most important—our fuel and raw materials. But after consuming inputs, we have to use them to create a complete cycle of growth. The right inputs now require the right actions to fully leverage their benefits.

Taking consistent action towards what we want is by no means easy; if it were, everyone would do it. We can assume that whatever outputs we want will require us to level up our current abilities. In order to achieve these outputs, we'll be forced out of many comfort zones. In order to thrive while forcing ourselves out of comfort zones, we'll need to take the chemistry of our brains into account.

The first thing to understand about the actions we take is that they are determined by our willpower. The only way we are able to consistently exert ourselves and do hard things is by understanding willpower. In her book *Willpower,* scientist Kelly McGonigal, PhD, outlines willpower as, "...the ability to do what you need to do, even if part of you doesn't want to."[13]

The best outcomes we want to create lie behind what part of us does not want to do. We must regulate, build, and safeguard our willpower through our inputs and actions. Anything which stresses us out or makes us feel overwhelmed degrades and tests our willpower. The easy answer is to limit our exposure to these things. But the reality of it is that this isn't so easy, and in the near term, there are many things outside our control. We'll always encounter stressors, which is a good thing. The best things in life wouldn't be worth anything if they didn't lie behind huge stressors. We value outcomes proportionally to the amount of willpower required to achieve them.

In order to boost our willpower, we must strategically regulate the raw materials we consume and the actions we take. If we choose to, we can fully control these parts of the equation: diet, sleep, and exercise. By taking the action to regulate these, we preemptively build our willpower for whatever fate brings our way.

By focusing on these three things—in this order—we create a self-reinforcing cycle which helps our bodies create and "endogenously dose" our systems with powerful drugs. These powerful drugs are neurotransmitters like serotonin, epinephrine, nor epinephrine, dopamine, testosterone or estrogen, human growth hormone (HGH), and brain-derived neurotrophic factor (BDNF).

These drugs all enable us to achieve the outcomes we want. They reinforce, build, and protect our willpower. By regulating our diet, sleep, and exercise we dose our brains with the right drugs to feel good about ourselves in a sustainable way. When we feel good about ourselves, we want to take care of ourselves. This makes it easy to exercise. When it's easy for us to exercise, we push ourselves harder and get the full benefit of a workout. When we work out hard, we're able to sleep more easily. When we sleep more easily, more deeply, and more soundly, we bathe our brains and muscles with the HGH needed to properly recover from exercise. This cycle reinforces itself and helps our abilities and capabilities expand.

Exercise

Cardiovascular exercise has been shown to create brain-derived neurotrophic factor, or BDNF. Dr. John Ratey profiles the creation of this protein vital for brain health and growth in the book, *Spark*. The most useful way to remember why BDNF is so important to our brains is through the nickname Dr. Ratey gives it: Miracle Grow.[14]

We can't stroll down to the store and buy BDNF. Even if we could, adding things exogenously to our body comes with a host of unintended side-effects. When we begin mastering our bodies and exercise, we can create BDNF naturally. BDNF is the key to begin building and maintaining healthy brains. It's a relatively cheap super drug or medicine (depending on your view) we can access easily. BDNF requires we pay only a small price of our time with cardiovascular exercise. BDNF also facilitates the growth of fatty coating that insulates our neural pathways, called myelin.[15] Just like electrical wires with the proper insulation, when we learn something new, we can "insulate" the neural pathway involved with performing that task by endogenously releasing BDNF.

It's sad to watch so many people begin exercise or weight training, then quit or become injured before they see any results. Every time this happens, the root of the problem is their inputs and actions aren't in alignment. When we stack diet with the actions of exercise and sleep, our bodies and brains are rewarded with human growth hormone. When we don't stack this trifecta together, we are exposed to injury and burn out.

Interestingly enough, the sleep part of the equation is the most difficult to implement. So why is getting great sleep so difficult? We can define "great sleep" as when we get eight hours, with optimal time in stages three and four of rapid eye movement (REM). We don't need to worry about the stages of REM if we have our diet and exercise in order. These will help our bodies proliferate ghrelin and HGH secretion.[16] Ghrelin helps regulate appetite and cravings, so a full eight hours of sleep is mandatory if we're going to ever stay clean from sugar and gluten. The release of HGH is mandatory if we're ever going to recover from tearing our muscles and the stress responses associated with working out.

Getting adequate sleep is most difficult because of our culture. The more peculiar parts of our culture associate sleeping less with toughness. The person who is "tough" and refuses to get adequate sleep inevitably ends up as a miserable human being with erratic emotions and hormones. They suffer mood swings, become depressed, and this masochism leads to diabetes, heart disease, and eventually dementia.[17] The "toughest" among us who "don't need sleep" also volunteer for perpetual low levels of testosterone and estrogen. Noting how stupid cultural quirks are is the first step in becoming cultural alchemists who build and create culture, instead of consuming it. Our society bemoans suicide, yet seems to support incremental suicide via crappy diet, lack of exercise, and little sleep. It's sad that nobody speaks up more about these true root catalysts of depression and suicide.

The machismo of competing on who can get by on less sleep is a fools' game. It's especially deceiving because most of the costs of the go-without-sleep game aren't paid until later in life. The selfish part is that the costs of these illnesses are often paid for by our families or children. The most selfish thing we can do now is to put the future problem of our health onto our families later on.

Think about the stages of poor diet, sleep, and exercise. People who engage in it now make themselves and their families miserable in the near term. They fail to become a function which ever produces positive outcomes. Eventually they bring an illness onto themselves which forces others to take care of them. Those who refuse to take the actions of creating a clean diet, great sleep, and adequate exercise are not just stupid– they're downright selfish.

Making exercise and sleep into priorities which we enforce is vital to any outcome we want to create. The outcome we envision for ourselves has no chance of happening until we continually sell ourselves on how amazing that outcome could be. Chances are we will never be able to fully sell ourselves until we have the right incentives and stakes. We need people in our lives who want to achieve the same outcome, or we need to find a community online like Lift.do.

Whenever there is an outcome we want to create but have not yet become active in achieving, we only need to examine the incentives and stakes. By tweaking and adjusting these, we influence our likelihood of achieving the outcome we want.

When the right blend of incentives and stakes hits a tipping point, we will be forced to take massive action. Before we create the right blend of incentives and stakes, there is no amount of willpower or motivation that gets us to act. Experimenting with our inputs and actions by placing stakes and incentives around them is the only way to start engineering the outcomes we want. It's the only way to make it to that crucial 66 days (on average) it takes to form new habits.[18] Stakes and incentives are the only powers strong enough to change our behaviors in the most vital inputs and actions we have: diet, sleep, or exercise.

We've seen that diet, sleep, and exercise are the precursors to achievement. Now let's explore how to perfect, hack, and optimize them.

Strategy and Action

"As you know by now, antioxidants help control marauding free radicals...but we can't do this by consuming antioxidants themselves. Our DNA can actually turn on the production of protective antioxidants in the presence of specific signals, and this internal antioxidant system is far more powerful than any nutritional supplement." –Dr. David Perlmutter

"Our data support the hypothesis that the formation of... compounds generated from oxidation of EPA and DHA in vivo can reach concentrations high enough to induce Nrf2-based antioxidant and... detoxification defense systems." –Dr. Ling Gao

The Most Powerful Prescription

These are experiments and ideas to get us started and focused on the three most important actions we can take every single day. These three actions combine to give us access to a huge endogenous supply of performance enhancing drugs: HGH and BDNF.

One Day at a Time

Hang a year calendar on the wall. On days when we've gotten enough sleep, kept our diet clean, and exercised (even if it's walking), put a big check on that day on the calendar. The calendar has to go up in a place where we're forced to look at it daily. Seeing all 365 days in the year is a powerful reminder that the most important battle to win begins every single day.

Exercise

Cardiovascular exercise to dose BDNF:

Running: Wherever possible, run and sprint. Try to do this off of pavement to prevent injury. Find an area with soft grass, or take up swimming. Running or walking on pavement is preferable to doing nothing at all, but running on grass or swimming is much, much better for the long term.

Rest Days: When we work out hard and smart, we only need to do it three days a week.

More ideas for creating incentives and stakes around exercising:

Martial Arts: Get two to three friends and go to a private center (off base if you're still in the military). Avoid blows to the head because of the whole traumatic brain injury risk.

Yoga: Find a class on Meetup.com or at the local gym.

If there is a Soldierfit or veteran-owned gym nearby, register for a group class. http://www.soldierfit.com

Bikram Yoga: This is an experience which, when done right, leaves most people with several hours or an entire day of calm, sometimes euphoric after effects.

A class can be helpful because at the end of the day, when our willpower is drained, something we've already paid for (incentive to go) where class members or friends are counting on meeting us there (stakes) is easier to force ourselves to attend. If we find ourselves not doing these activities, we only need to ratchet up the stakes and incentives until we act.

More Ideas for Exercise Stakes and Incentives

- Sign up for a workout class with a friend or spouse who motivates us.
- Bet money on whether we will or won't accomplish exercise with somebody who underestimates us.
- Do them with a group so we can't back out.
- Pay money for classes which are nonrefundable.
- Realize that getting up and going to the gym requires more willpower than actually working out once we're there. If we can just make a few taps on our phone to call an Uber, we've already instantly applied stakes and incentives to grease the wheels forcing us out the door and to the gym.

Sleep

The biggest threats to sleep we face? Alcohol. Sugar. Gluten. Alcohol in our bloodstream is instantly converted to sugar. It completely destroys our sleep and REM time. I've found any time I drink, even in the afternoon, my sleep suffers. Drinking after five o'clock just isn't as helpful in comparison to having one or two drinks in the early afternoon.

Another big threat? Never catching up on sleep we've previously missed. This is tough to accomplish, but it's important to do our best to get caught up.

Looking for a catch up short cut? Try spending an hour in a float tank. If we're severely lacking in sleep, or have spent years without enough sleep, consider "catching up" with a float tank. Float tanks, on average, give us the equivalent of four hours of sleep for every hour spent in the tank. After a certain point these returns will be diminishing, so please don't get in a float tank for twelve hours. Go to a local professional provider with a new, well filtered and cared for tank. Float tanks take some getting used to, but the benefits from them are enormous. Expect them to completely sweep the country in popularity within a few years. In a few years, float tanks will be seen as a must have tool for accelerated meditation, healing, and general well-being.

Additional ideas and strategies for amazing sleep:

- Block out all the light coming into our rooms
- Breathe Right nasal strips
- White noise (can be a fan)
- Optimal room temperature for sleep: 60-66 degrees, this is important to test and see what we respond to.
- Don't sleep with a cell phone, electronics, or TV in the bedroom.
- No screens after 6:00pm.
- No caffeine after 10:00am.
- No big meals after 7:00pm.
- No alcohol after 5:00pm.
- If we miss sleep, make sure we catch up.

Optimal amount of sleep to shoot for: seven to nine hours. Once we're completely caught up on sleep, and optimize our diet, we may find ourselves needing progressively less sleep.

Test this out. Record it if necessary. As we go, examine all the cultural silliness around the taboo of sleeping enough. Examine how worried people become when they say, "I don't need sleep." These hard chargers eventually find themselves with an inability to charge, literally.

8

The Mental Game: PTSD or PTG?

"We hear about the more lurid post-traumatic disorder, not post-traumatic growth, in the intellectual and so-called learned vocabulary... Intellectuals tend to focus on negative responses from randomness (fragility) rather than the positive ones (antifragility). This is not just in psychology: it prevails across the board."
–Nassim Taleb

"Relieving the states that make life miserable...has made building the states that make life worth living less of a priority. The time has finally arrived for a science that seeks to understand positive emotion, build strength and virtue, and provide guideposts for finding what Aristotle called the "good life." –Dr. Martin Seligman

At a Bar Overlooking the Puget Sound

"So *why* is that the goal?" I asked.

The man I was asking had been a psychiatrist for 20 years. He was growing irritated... and what appeared to be uncomfortable. His eyes squinted and, for an instant, I detected disgust on his face.

He was making it painfully clear he did not appreciate questions; especially from someone *like me*. He let out a sigh, and then answered.

"The goal is to get patients with PTSD back to where they were because it's what's best. It's a long and *sophisticated* process," Dr. Luddite said. His eyes squinted at me for that extra measure of hostility. I couldn't help but wonder if he was trying to look tough or constipated.

"I'm sure that it is sophisticated. My point is, why is getting back to average the pursuit? Shouldn't there be an obsession with finding post traumatic growth?"

This guy was less frustrated now, and he responded, "It's not that simple."

I understood part of his hostility. I was almost 20 years younger than he, and it had clearly been a long time since he'd ever had to answer any questions. But we were both speaking at the same event in the Pacific Northwest. The speakers were all meeting beforehand at a happy hour, and the venue was gorgeous. We stood overlooking the Puget Sound. I had a suspicion that I wouldn't become friends with this guy anytime soon—but I genuinely wanted answers from the front lines of psychology and psychiatry. These professionals were interacting with so many veterans, and I had serious doubts about how practitioners were approaching the issues.

"I think in the coming years, practitioners who work with veterans will have to show results. That's because the veterans paying for help are going to only want to work with those who set post traumatic growth as a goal. I think the whole treatment of trauma is going to get flipped," I said.

"Well… things are changing, but it'll remain our job to tell them the appropriate goal based on where they're at. How is the Internet going to change that?" answered Dr. Luddite.

"People aren't going to pay to get back to normal, because that's impossible. People are only going to go to psychologists and psychiatrists with the attitude that post traumatic growth is the goal. They make their decision on their phones or computers, and they want proof—they want testimonials. Word has gotten out. Not only do veterans not have to carry around the disorder label of PTSD, but if others have achieved PTG, they'll want that too," I said.

Dr. Luddite made some sort of grunting noise and took another sip of his drink, "And you really think they'll be *demanding* professionals like myself to comply with what they want?"

"Demanding? As a customer, don't they get to ask for a solution?" I responded.

He gave a creepy response hinting at the fact that he would still find a way to get paid. The sad part is, this guy didn't understand how business works or the fact that his patients weren't asking for much because they weren't paying him directly. As long as they had a basic amount of faith in that industry, and their insurance picked up the bill, he was right that he would continue to receive a stream of patients.

"I'm willing to bet that the Internet will change everything. It's not being viewed as *weak* anymore if you talk to someone. Eventually, the only things vets want are cures and answers. Those who provide the best medicine and solutions will have a line out their doors."

"Name a single professional whose focus is on post traumatic growth," demanded Dr. Luddite.

"MAPS and Anthony Robbins come to mind (Multidisciplinary Association for Psychedelic Studies)." I smiled.

"I don't know what MAPS is... but Tony Robbins?" The psychologist smirked with laughter, "isn't he an infomercial guy?"

"Yes, that Tony Robbins, the one that billionaires and Presidents hire for advice. I think he's one of the forces that helped you guys finally find positive psychology. Don't get me wrong, Seligman's work was crucial. And of course, Victor Frankl was a master of alchemizing post traumatic stress into growth. Oh andAnd MAPS is a group helping fund, oversee, research, but generally they're curing traumatic stress symptoms after just three therapy sessions. That's a bit quicker than the whole CBT and SSRI routine," I said.

"Mmm hmmm, well I'm still not familiar with MAPS, but Dr. Seligman is actually a *doctor and researcher*, you see," said Dr. Luddite.

"I think a track record of successful results are more important than titles, don't you? I asked.

"Successful? I think your definition of success might be a bit of a stretch if you're talking about the people he works with..." Dr. Luddite smiled as he took a long sip of white wine.

"Well, if people pay $3,350 for a Robbins seminar, he's doing something right. He's worked with three Presidents, actors, professional athletes, Silicon Valley entrepreneurs, billionaires, even Oprah. Oh and another practitioner in your field just started doing the same thing, Dr. Ellen Langer. She does her seminars and retreats and charges for them because it works really well," I said.

"I don't see what the super rich have to do with piecing broken men and women back together who are suffering from PTSD," said Dr. Luddite.

"I'm just saying that veterans are going to increasingly work only with those whose focus is on helping them find post traumatic growth. Organizations like MAPS are curing veterans of stress symptoms in just a handful of sessions, plus they're getting real video testimonials from those they work with. It's only a matter of time before vets will pick those who prove their cure with video testimonials." I was going to keep going, but could tell Dr. Luddite was too basic for even a fun argument.

Dr. Luddite mumbled and walked away. I took a sip of my drink and looked out at the Puget Sound and smiled. Thank God for the Internet.

The Mental Game

So what comes after fueling our bodies with healthy raw materials, taking action to learn each day, exercising, and sleep? The only thing left in the health part of the equation is the biggest part—the mental game. By monitoring our diet and regulating our willpower through exercise, sleep, and learning daily, we've primed ourselves to win the mental part of the game.

Disclaimer and Warning

Skip this section if you don't need it or aren't lawsuit-happy.

Of course I can't know for certain where every individual is at, and it's up to each of us to, when in doubt, reach out and ask for help. There is no shame in seeking to heal or optimize ourselves. That takes far more courage in our culture then many realize. Whenever we need to talk to someone, we should feel free to do so. There are plenty of free and amazing sources, like Military Chaplains who are on duty 24/7 at 1-800-80-VETS-4. This chapter is an alternate view on the future of mental health for veterans and anyone facing symptoms from post traumatic stress.

Like everything in this book, my remarks about post traumatic stress (PTS) are my opinion. I'm not suggesting my prescriptions be applied across the board. These are all experiments which can be tested or discarded. This book is about sparking better ideas for individuals. PTS, TBI (traumatic brain injuries), and every other mental challenge veterans might face are understandably hot-button issues. I'm not a doctor, psychologist, or psychiatrist, so take that into consideration before you read on. But also take into consideration that the only reason I write that is because parts of our culture have de-evolved into preferring outrage and censorship to rational discussion and logical arguments. Other parts of our culture have de-evolved to prefer suing people over imagined transgressions, or slandering them because of a disagreement. In a funny way, lawsuits have become a type of lottery for pseudo-intellectual trashy people. Virtuous people aspire to start a business and become wealthy, trashy people scheme lawsuits.

A New Paradigm on Government and Dropping the Disorder

PTS, TBI, and the host of challenges veterans face have been mishandled since the start of our conflicts overseas. The military has no successful track record of getting veterans proper or beneficial care after their deployments, and I don't expect them to anytime soon. Should they provide better care that focuses on the root problems instead of symptoms? Absolutely. Do I think wasting time arguing about it or citing the specifics is useful? Not at all. Will it help to write more about the 600,000 backlogged claims from veterans trying to claim their benefits? Maybe a little. But I'd like to offer a different approach, with different prescriptions that every veteran can take advantage of immediately.

My philosophy is simple: Anything the government promises me, I don't count on receiving. Any type of training, therapy, or medication the government provides through the VA might be helpful or harmful. I don't know the specifics of everyone's individual situation, but I do know the most useful way to view what the government doles out: there is a better alternative in the private sector. Almost every single thing the government offers can be done more efficiently by the private sector. This wasn't always true, take NASA for instance, they sent us off into space, but quickly fell into obsolesce. Life is too short to look to a drunk, indebted uncle for solutions to our problems.

The second part of my philosophy: There are plenty of people in leadership positions in government and medicine who have sat home in comfort and taken no personal risks (or extremely limited risks) for the economy or our country. These types of people are the ones who have coined the label, PTSD. Within the total population of people trying to draw awareness to the challenges veterans face, there are many with good intentions. But I'm not going to accept the label of "disorder" from these people, and I don't understand why veterans and others choose to use the phrase. The only thing these awareness campaigns have done is spread the idea that something is "wrong" with the average veteran.

The common sense approach to viewing the reaction to extreme, traumatic events is simple to me: a stress reaction to trauma means our bodies are working. This is our body and brain signaling to us that they're functioning properly. I'm not saying the events which caused the stress response were good—they were horrible. But the fact that our bodies are signaling that they've been through too much stress is a great thing for us. I view this only as a signal. I don't even like to use the word "symptom." It's a shitty mental state to be in, and there is plenty of help to get better.

The best chance veterans have to thrive is through seeking out practitioners, therapists, and psychiatrists who are only interested in completely healing them. The people veterans decide to work with must be on the cutting edge of their fields, and veterans should accept nothing less.

If the lessons of trauma and adversity throughout history have taught us anything, it's this: trauma is horrible—but it's an opportunity to become something more.

"The more challenging or threatening the situation or context to be assimilated and affirmed, the greater the stature of the person who can achieve it. The demon that you can swallow gives you its power, and the greater life's pain, the greater life's reply."
– Joseph Campbell

There is no greater victory than becoming an incredibly high achiever in the face of all those who just expect us to have a disorder. We should embrace whatever "disorders" we have, and aggressively study those who have used similar disorders and similar pain to find prosperity. The litany list of these people is long. The list of practitioners who would love to work with veterans this ambitious is long as well. There is no better way for veterans to "win" than to become wildly healthy, self aware, skilled, rich, and successful. While the government dumps billions of dollars down the black hole of the VA, spreading the word to employers that we all have "PTSD," we can smile and know otherwise.

When our bodies produce a natural biological stress reaction to extreme events, and we get a cognitive response and symptoms of it, this should be a cause to say, "I'm glad my body is working okay," and not, "I have a disorder." Telling people they have a disorder and that you're the only one able to alleviate that disorder is an age-old sleazy sales technique. If you're an evil person and you want a steady stream of money for life, you simply convince children that just by breathing, they are inherently evil. Then offer a promise of alleviating their guilt via payments, and you have a steady stream of income for life.

Teaching people to self-examine and self-attack for specific negative behaviors which only you can treat is basic evil. We cannot succumb to these evil institutions.

The dirty little secret about the lives of most Americans today is that many are afflicted with diseases of comfort. Adversity is the secret to success, and veterans have an unfair advantage in a society where many are crippled by comfort.

Post Traumatic Growth, and the Power Behind Adversity

Every adversity that we face—every obstacle, challenge, attack from an enemy—holds the key to an equivalent advantage. Sometimes the disruption and pain of adversity is so great that it's only after years and with the help of a coach that we're able to fully realize the value and scope of the equivalent advantage.

Psychologists and psychiatrists have named the appearance of this phenomenon "post traumatic growth." What's a bit silly to me is that if post traumatic growth is accessible to anyone who faces a challenge or stress, why isn't fostering that the sole goal of treatment? It is the sole goal in treatment, but only from a subset of psychologists and psychiatrists, not from the mainstream community.

As I mentioned before, I think it's important that the treatment provider we select have the expressed goal of not just treating the symptoms, but helping heal us in a complete way. Most treatment providers for PTSD start with prescribing SSRIs. Simply trusting a "provider" to administer treatment is a recipe for trouble. There is a second level of research we each have to conduct if we're considering pharmaceutical treatment from psychiatrists for symptoms we're experiencing. Like all things in life, trust but verify.

The Future of Medicines

"The VA is an incredible black hole. You won't be able to fix it unless you shut it down and rebuild." –Marcus Luttrell

So by now, we've probably all noticed that the Veterans Affairs *is not* receiving glowing video testimonials from veterans. No veterans are raving on social media about how the VA has helped them.

The far better questions to start asking are which individuals and organizations have the goal for veterans to achieve post-traumatic growth? Are there any organizations or practitioners out there who are healing veterans in short amounts of time? The answer to those questions is yes, there are organizations, individuals, and medicine responsible for completely healing symptoms of post-traumatic stress and even traumatic brain injury.

Some of these organizations include John Hopkins University, MAPS, New York University, Los Angeles Biomedical Research Institute, the Imperial College of London, the University of Zurich, and UCLA. They are all working to heal those with symptoms of post traumatic stress, anxiety, autism, and addiction. The medicines they're using include hemp, psilocybin, and MDMA.

They have testimonials and results from their initial tests, which are nothing short of amazing. The medicines they're using are, of course, nothing close to drugs. MAPS and the universities running these experiments have received the necessary approvals to use them in therapeutic sessions. These researchers need brave veterans for future cohorts and tests. Medicine and human knowledge of medicine cannot progress until risk takers pursue opportunity. MAPS sees a large opportunity to help speed peace of mind and healing to those who need it most. Contact details and other information can all be found at http://www.maps.org/research/mdma/ .

It would be stupid, of course, to expose ourselves to the risk of trying to experiment on our own, but the best psychologists and psychiatrists will be knowledgeable about these studies. When in doubt, ask a professional, and don't settle until you've found the best.

The future of medicine doesn't lie in drugs, it lies in blending the best research we've done with the best medicines which have already been perfected from 4.5 billion years of evolution. Many geneticists and biologists would argue that large portions of what we know as "illegal drugs" today will be known as medicines in the near future. Let's consider the substances these studies used to treat veterans: psilocybin, hemp, and MDMA. I use those words strategically, as the other "names" that much of the media uses for these "illegal drugs" are callous attempts at either poisoning the well, or a general ignorance of where the slang names for these substances originated.

MAPS estimates that it will take the FDA at least five more years to gain approval for using MDMA. At a certain point, we might all have to admit the FDA is severely broken, and the estimated 7.7 million Americans who have symptoms of traumatic stress should be free to choose the best treatments available.

I would encourage anyone who is skeptical about the results of these studies to reach out to the participants. Or just do a few quick Google searches for psychedelics and PTSD therapy. There are an increasing number of video testimonials of veterans and non-veterans who have used these medicines in legal, safe settings with therapists. The results are astonishing, and so far many studies have had over 80% of participants cured of symptoms after only a single, or couple treatments.[19] Many of these patients were previously debilitated by their symptoms. These video testimonials are astonishing, and should seriously worry any therapists, psychologists, or psychiatrists who don't want to completely heal their patients.

For those who feel supporting this medicine and therapy is important, there are many ways to participate, contribute, or join these studies. It is as simple as going to Maps.org, signing up for the Maps.org newsletter and staying apprised on their updates. Maps.org has a research tab on their website which contains links to all the current, concluded, and upcoming studies. Participating or helping is as easy as finding the names of those conducting the studies and reaching out.[20]

The crazy thing about Maps.org is that they're fighting a huge battle just to gain the approval required to treat people with these various "illegal substances." Whatever your take is on the war on drugs, it's important to recognize that these substances are cheap, abundant, and are producing amazing results for service members. These substances are treating in moments what some people have suffered quietly with their whole lives. Thanks to the laws and aftermath of the war on drugs, trying to pursue any of these treatment methods on our own is asking ourselves and our families to take huge unnecessary risks. Maps.org has already done the work for us in many regards. We can let them handle the legal issues for us, and we can get all the information about how to safely apply and receive treatment to completely cure many of the challenges we face.

Simple, Safe, Effective, and Free

There are plenty more steps we can take to better understand or deal with any past trauma. It's better, of course, to work with a professional, but while we wait for appointments (or in the case of the VA, you might be waiting months) there are simple and free practices we can begin. Meditation and mindfulness can be started right away. When coupled with coaching on our own terms, these can become powerful practices.

Strategy and Action

"Never give in. Never give in. Never, never, never, never—in nothing, great or small, large or petty—never give in, except to convictions of honour and good sense. Never yield to force. Never yield to the apparently overwhelming might of the enemy." –Winston Churchill

"Some things benefit from shocks; they thrive and grow when exposed to volatility, randomness, disorder, and stressors and love adventure, risk, and uncertainty. Yet, in spite of the ubiquity of the phenomenon, there is no word for the exact opposite of fragile. Let us call it antifragile." –Nassim Taleb

It's Not "Therapy," It's Coaching

The *Headstrong Project* is the best traditional route I can recommend for any veteran who faces anxiety, depression, PTS, TBI, anger, addiction, or sleep disorders. The *Headstrong Project* is led by Zach Iscol, a Marine Officer, Iraq veteran and entrepreneur. "The Headstrong Project is a non-profit organization that provides cost-free and bureaucracy-free mental health care treatment to Iraq and Afghanistan." http://getheadstrong.org

Code Words for Therapy

In business, technology, and entrepreneurship, there is a code word for therapists; they're called coaches. Top performers hire coaches for help in every aspect of their lives, plain and simple. The point is, we should never let any negative associations around therapy or coaching stop us from reaching out to experts. If we think of this as "getting help," it can be debilitating psychologically. It's not getting help as much as it is investing in ourselves. It's not admitting weakness to seek out an expert.

The most important thing about this is that it's completely accepted in the top echelons of the business world. Once again, thanks to many of the positive shifts in our culture, Silicon Valley has made finding a coach (instead of a therapist) just a part of the weekly routine.

There's no shame in becoming the best version of ourselves, by any means possible, which includes liberally hiring therapists or coaches— and firing them as we see fit. I've hired them and I've fired them, and I will continue doing so. The best ones have been an amazing investment, and there are few things as satisfying as firing the bad ones.

My point is simple: we have nothing to lose by reaching out and talking to someone. The people at the top get coaching. Some call it therapy and some call it coaching. The best of the best do it, and we can too. When done right, it can be a gateway to self- knowledge and a huge relief from anxiety. Many hugely successful CEO's have "coaches" who follow them around.

Depression and mental struggles are slowly being talked about openly, and the technology community is leading the charge. Depression is a hallmark of creatives; its nickname is the "entrepreneur's disease." If there are experiences we've had that we can't stand to think about, therapy and coaching exist for us to force those experiences to serve us. The best at their craft will help us find the equivalent advantage in that adversity. It might sound insane at first, but the best will help us find this and become something more.

Talking to someone is as simple as calling one of the chaplains on duty 24/7 365 days a year: 1 (800) 80-VETS-4

Or if you don't want to talk to a military person, get paired up with a civilian counselor 24/7 365 days a year: 1-800-273-8255

Or we can go to the Head Strong Project or Maps.org and ask them to recommend someone in our area.

No Shame in Being Human

"I am a human being, therefore nothing human is foreign to me." –Publics Terentius

Therapy is just now starting to be discussed openly among entrepreneurs like Noah Kagan, Brad Feld, Tim Ferriss, Rand Fishkin and many more. Many successful entrepreneurs see therapists behind the scenes. Those that do it aren't admitting weakness as much as they are admitting a willingness to level up.

What I look for when talking to a therapist or coach is somebody who:

- Is entrepreneurial
- Is seeking the same virtues, and holds values similar to my own
- Has achieved things I respect

Mindfulness

"But let me explain to you that it's the culture that teaches us that we have no control." –Dr. Ellen Langer

The oldest and most widely used way to transform stress into benefit is through mindfulness. Some people call it meditation. Great religious teachers have referred to it as "being present" or being "in prayer and worship at every moment." What they are all talking about is simply noticing what goes on in our minds and recognizing the difference between useful and not useful thoughts. The thoughts, which rise and flood over us second by second, are not us. They are just thoughts—possibilities. When we step back to observe them, we get the power which comes from recognizing entrenched mental habits and patterns, our brains repetitively churning out worry, stress, and fear. Mindfulness begins when we take each thought and worry which arises and watch it, then label it as useful or not useful. That's it. No teacher or guru needed, although a teacher can very helpful because this practice is much easier said than done.

When I first considered meditating, I imagined sitting down and humming or chanting. This, of course, isn't what meditation is. Mediation can be guided, but it's really just sitting or lying down, being quiet, observing our thoughts, and focusing on our breathing.

More and more CEOs and incredibly successful people, people who have overcome adversity like Steve Jobs, Oprah Winfrey, Rupert Murdoch, Mark Bertolini, John Mackey, Rick Rubin, Jeff Weiner, Soledad O'Brian, Russell Simmons, Arianna Huffington and many, many more have used meditation and cite it as the primary thing that allowed them to maintain the mindset necessary to push past every obstacle to get what they want.

Tools, Apps, and Hardware for Better Meditation

Apps to use to begin meditation and mindfulness practice:

Headspace: "Meditation made simple" https://www.headspace.com

Lift: a habit-tracking app. Great to find a community of others to "lift" us up to meditating or practicing mindfulness. https://www.lift.do

Muse: For more guidance, tracking, and proof that we're doing meditation "right," I highly recommend the Muse headband. It's still early days for this technology (BCI), but the results look very promising. The headband combines with the Muse app to help us monitor, measure, and quantify our meditation. http://www.choosemuse.com

Old Faithful: This is one of my favorites. Get a year-long calendar where we can check off every day we've meditated. We can use it to track whatever we're going for, whether its getting enough sleep, eating right, exercising, anything. It works great when we simply know what we want to accomplish each day. Once we accomplish it, place a little check mark on the day. I use a simple, 365-day calendar from a small design shop called BBT.[21]

Part 3: Relationships

9

The New CEO

"For though we love both the truth and our friends, piety requires us to honor the truth first." –Aristotle

"I've always thought that each person invented himself, for whatever reasons, through whatever circumstance, through whatever he has gone through, that we are each a figment of our own imagination." –David Geffen

In the Middle of a Startup

"Get out of here!" cried Steven.

"What do you mean? I'm just saying that we're not gonna make more sales until Jonas gets us better leads," answered Danny.

"It's not what you said, it's how you've been saying it. It's how you've been thinking, and it's poisonous," said Steven.

Steven was only thirty years old and already worth a bit over a million dollars. We were sitting in the office of his company which he was self funding. Danny was one of the first salespeople we had hired. The week after we hired Danny, we started to detect some occasional irritability from him. And now our suspicions had been confirmed. When Danny was angry or frustrated, he let his emotions spill out to the rest of our small team. His anxieties were spreading to the rest of us. We were in the fragile early stages of a startup and were barely making enough money to keep the lights on.

"I'm dead serious, Danny. Leave. You're fired!" Steven said.

His words fell onto the ears of someone who was completely blind to his own emotions. Danny started to try to say something to object, but now Steven was standing, pointing to the door, yelling, "If you ever want to do anything in life that matters, you have to learn to control your emotions. You either bring solutions to the table, or you're only spreading problems. I don't have the extra cash or time for you to learn about attitude here. Learn it someplace else."

Danny rolled his eyes and sulked on his way out the office door. The sad part was that he didn't realize the truth in what Steven was saying. It was so easy for one person to poison the emotional wellbeing of a tightly- knit team. Bringing anything less than full control of his emotions into the group was harmful. It was fine to bring up worries and risks, but they'd better be paired with ideas and prescriptions on how to mitigate them.

We've all met or worked with people like Danny, who spread negative mindsets into their relationships. People like Danny don't realize that it's no accident that people like Steven start their own companies. There were reasons why Steven was obsessed over being positive every day. He got to where he was by learning to master his emotions.

In startups and businesses, the CEO is responsible for the culture. They do not bring in employees without thoroughly vetting them. In the early stages of a company, the most vital job is to set the tone, and only bring in those who fit emotionally with the rest of the team. The

CEO sets this cultural tone. Either they completely control their emotions as they flood to the rest of the team, or they harbor or spread emotional viruses. Those who don't focus on the emotions they create and spread, inevitably end up spreading viruses.

The CEO acronym has traditionally stood for the Chief Executive Officer. But the world we live in is being transformed by technology. Businesses and institutions wanting to innovate are increasingly relying on flatter organizational structures. This means successful CEO's must rely less on management, and more on leadership. They're relying less on top down control, and more on distributed teams working with higher degrees of autonomy. In order to be a leader in these flatter organizational structures, emotional mastery is required. We see a focus on this concept everywhere from Silicon Valley technology giants to Seal Teams as outlined by Mark Divine in the book, *Unbeatable Mind.* Those who succeed in life and their relationships are the ones who build their emotional intelligence.

From CEO to *CEO*

A more useful variant of the CEO acronym is Chief Emotions Officer. This acronym comes from Chip Conley, author of *Emotional Equations* and founder of Joie de Vivre Hospitality (a massive California luxury hotel chain).

If the phrase Chief Emotions Officer sounds corny at first, don't worry; I thought so, too. But it's often corny phrases and acronyms which stick in our mind, available to be recalled at just the right time.

The Chief Emotions Officer acronym is the same foundational approach a growing number of founders from multi-million to billion dollar companies use. Marc Benioff, the founder of Salesforce.com (market capitalization: $37.74 B) is a true EQ master. In his book *Behind the Cloud*, Benioff reveals over 111 "plays" which helped lead to the creation and sustainment of Salesforce.com. All these plays require emotional intelligence to execute. In fact, Benioff says the leadership model he used to build Salesforce.com came from Tony Robbins,[22] who is a fierce advocate of emotional mastery. As we saw earlier, founders often don't hire therapists, but they hire coaches to help them build up their emotional intelligence.

Many of the "support" and executive roles at technology companies exist to help founders increase their emotional awareness and mastery. Consider Sheryl Sandberg's success as Facebook's COO. She was there to help run the show, but also coach Zuckerberg and help him become a better-rounded CEO. Just like the best of the best hire coaches, they also focus on mastering their emotions.

Approaching all our relationships in business and personal life with the CEO mindset is transformative. This mindset is a habit built by intense engagement and presence in every emotional interaction, every time. Over time, the conscious effort and willpower necessary to implement emotional involvement becomes second nature.

The CEO practice can begin in many ways. Often it starts by catching complaints before they leave our mouths. Just by observing the negative emotions we feel, we gain an opportunity to conduct damage control, or subvert them accordingly. The next time a negative thought comes up it can either serve us by pointing us toward a solution, or we can discard it, as it's not useful to anyone. A CEO doesn't dwell on useless thoughts. Either they can be used, or they're noise or toxic complaints. There is rarely a middle ground.

In part I of this book we saw how starting with the foundation of health helps dictate what our own emotions will be. The mentality of treating ourselves as a function and monitoring our inputs gives us a healthy balance of brain chemicals. No sugar and no gluten means far fewer mood swings, and far more willpower for heightened states of emotional control and engagement.

So what are some tactical ways to become better at harnessing the CEO mindset? First, it's helpful to recognize that the CEO mindset is based on the emotional quotient (EQ) of our intelligence. As we raise our EQ, we'll naturally gain more control over our emotions.

In recent years, a plethora of research has found that it's the emotional quotient of our intelligence (not IQ, and SAT) which translates into success and accomplishment in life. The most compelling research on the importance of emotional intelligence comes from Daniel Goleman. His book, *Emotional Intelligence,* presents the case for the value of emotional intelligence.

To sum up and simplify much of his research, Goleman found that up to 66% of a leader's effectiveness stems from his emotional intelligence. To put it more simply, emotional intelligence is twice as important as standard measures of intelligence such as IQ, for predicting accomplishments, success, and happiness.[23] This means that EQ is the area of intelligence where our focused energies to learn deliver the most tangible results, money, and accomplishments.

How to Build Emotional Intelligence

"Successful people ask better questions, and as a result they get better answers." –Anthony Robbins

The quickest and most sustainable way to build our emotional intelligence starts when we become curious about everything. Curious people don't attack others. They can't hurt others if they ask questions. Curiosity leads to progressively better questions, and as long as we listen, develops our emotional intelligence. When in doubt or when we want simplicity? Ask better questions.

To dive more in depth, let's examine even more questions and traits we can use to build our emotional intelligence. Goleman's research shows that emotional intelligence is built from recognizing five core skills: self-awareness, self-regulation, motivation, empathy, and social skills.[24] Those skills are described below, along with questions to help us analyze, monitor, and develop each. Try taking the questions below and answering them, or using them to formulate your own.

Self Awareness

Analyze: Who are we? What do we stand for? This is simple, but most people don't ever take the time to think about it, or make choices based on the answers.

Monitor: How are we perceived by others? How have our emotions affected others in the past? How are they affecting them right now?

Develop: It's easy to fool ourselves; it's much more difficult to ask the people closest to us how we come across. Do we laugh at ourselves? Can we remain confident, yet still posses the ability to not take ourselves too seriously?

Self-Regulation

Analyze: Can we control our emotions? Do we lose control of them easily? Are we able to stop negative emotions in ourselves before they spread to others?

Monitor: Do our choices place the proper value on ourselves? How do the emotions we feel either serve us or hurt us? How do we identify a time when we were emotionally positive and replicate that?

Develop: We can practice mindfulness when watching emotions as they arrive by filtering them into two categories. The two categories our emotions fall into are either useful or not useful. Once we identify them, we have to transform or let go of the lesser emotions predicated on things such as fear. The more positive emotions such as thoughtfulness, hope, or creativity we can articulate and spread.

Motivation

Analyze: Do we keep our own motivation high through monitoring, controlling, and selecting our inputs? Are we aiming high? Do we want to, as Horace Mann said, "Win some sort of victory for humanity before we die." Or do we just want to scrape by?

Monitor: What patterns do our emotions follow? What are the things we do when our motivation is ridiculously high? What are we consuming? What are we producing? Who are we around? Once we have a good idea of how to control our own motivation, we can replicate or engineer the variables, inputs, and actions which take us there.

Develop: Do we consistently guide our thoughts to solution-oriented ideas? Do we surround ourselves with motivated people? Either we become experts at engineering our own motivation endogenously, or we'll always have to rely on exogenous sources.

Empathy

Analyze: Do we seek to understand other people? Do we listen well? Are we careful, curious, and empathetic when someone displays vulnerability to us? Do we seek to reciprocate empathy after others have shown empathy to us?

Monitor: How do we treat others when they demonstrate the courage required to show vulnerability? Do we listen well? Are we afraid of connecting emotionally with others? When we do connect with them, how does it feel?

Develop: How can we practice becoming better listeners? Can we practice saying things which showcase vulnerability and require courage? How can we demonstrate to others that we have listened to them when they speak? How can we better recognize when others tell us something which requires courage to say?

Social Skill

Analyze: How do we handle and navigate the existing relationships we have? Which of our relationships are mutually beneficial? Which ones are not? How do we find more relationships which make us feel alive?

Monitor: When have we lifted others up? In our relationships, how do we create or add value? How can we increase the value and joy we bring when we interact with others?

Develop: How can we better connect with others? How can we relate ourselves to others so our interactions require less explanation, and more spontaneity? When moments of connection happen, do we study what went into creating that moment?

These core skills give the mindset of a CEO. There is a lifetime of enjoyment to be gained from properly valuing and forging long-term relationships. But until we get into the habit of practicing these five skills, it's easy to pollute our relationships. Just by being aware of these skills and the CEO acronym, we plant the idea in our subconscious mind for later use.

A quick note about relationships which are negative, toxic, or leave us feeling drained. They're too much work, and life is too short to keep them. We should do what we can to immediately distance ourselves from them today. If there are relationships we have where we're not sure if they are healthy or harmful, the CEO practice helps us reveal their true nature. When we build our emotional intelligence, we instantly recognize others who don't value emotional connections or long-term relationships.

We all know that deployments bring out the true nature of relationships. We've probably all seen many people continue relationships which are only a burden for both parties. Reflecting on those dysfunctional and damaging relationships we've seen in the military is one of the best ways to remind ourselves why emotional intelligence matters.

You Treat Your Relationship Like a Business?!

"For warriors in particular, if you calm your own mind and discern the inner minds of others, that may be called the foremost art of war."
–Shiba Yoshimasa

Some people have a hard time thinking about regulating their emotions. They think terms like "adding and creating value" sound too *businesslike*.

Speaking of relationships in an entrepreneurial fashion has a tendency to frustrate these types of people. The objectors say something along the lines of, "What if our significant other found out we're treating our relationship like a business!?"

This is usually an excuse to avoid doing the work necessary to build emotional intelligence. For argument's sake, what if we *were* treating our relationship in an entrepreneurial way?

All that means is we would have to bring positivity, creativity, humor, novelty, and fun to our interactions. It means we would have to practice vulnerability and honesty in those relationships. It means we would have to work hard in order to understand the other person's hopes, fears, and dreams.

Approaching everything in life with an entrepreneurial mindset, including relationships, business, learning, and health, is just plain smart.

Strategy and Action

"If I build really strong, true relationships…then all this other stuff I want to happen will happen." –Chris Sacca

"You are the average of the five people you spend the most time with." –Attributed to many

Relationships Audit

The idea of this strategy and action section is to get a better handle on where our emotional intelligence and relationships currently stand. The activities below are aimed to help us gather intelligence, create a plan of action for weak spots, and spotlight where we've already made progress.

Think of the two, three, or five people you spend the most time with right now.

What emotions do they most often bring to you?

1.
2.
3.

What emotions do you usually bring to them?

1.
2.
3.

We want to identify which of our relationships brings or asks us to carry an undue amount of emotional baggage. Drill down on that negative relationship even more; what specific negative emotions do they bring or elicit in you?

After we identify it, we want to figure out what the opposite or positive version of this emotion might be. It's the opposite of this emotion that we'll attempt to enact and mirror for the other person.

For instance, many people have the habit of coming home and detailing the worst part of their day to their spouse. This act is basically just dumping negative emotions onto someone like they're an emotional trash can. Once we identify this emotional dumping which makes us feel drained, we simply mirror the alternative, which might be sharing the best part of our day.

As a CEO, it becomes our job to mirror, cultivate, and engineer the emotions we want to see. So instead of receiving the emotional weight of the worst part of our spouse's day, and then reciprocating with the worst part of our own, we want to do the opposite.

When we get home, we talk about the best part of our day. We bring up and replay the positive emotions we felt when the best part of our day happened. Our spouse gets to feel those positive emotions, and we're one step closer to becoming a CEO.

When people consistently complain, fall short, or are emotionally weak, it's often in an area where they are hurting. There is almost always a deeper issue they're not dealing with. It's by displaying the healthy emotional alternative to their angst that we give them an invitation to connect on a deeper level. We also give them an opportunity to see how ridiculous they're being.

By reflecting and mirroring the emotions we'd most like to see, we position ourselves to receive them. We'll never be able to convince people to change their emotional state through an argument.

We can only change our own—and through our actions, give others an invitation to do the same.

The second part of the strategy and action section is to analyze the relationships we have with the five people who are closest to us.

List their names, followed by a quick summary of their hopes, dreams, and fears. These are the emotional core of a person. If we don't know these things about someone, we don't have an emotional connection with them.

Many people want to connect on a deep level emotionally, but have been hurt so badly in the past, they no longer risk letting others in on their hopes, dreams, or fears. See if we know the answers to those three questions for the five people closest to us:

Name	/ Hopes	/ Dreams	/ Fears
1.			
2.			
3.			
4.			
5.			

Now, we'll treat these results in a similar way to the first exercise. Where we don't know the hopes, dreams, and fears of the person closest to us, we will make sure they know ours. This can be extremely difficult to bring up, but it is an invitation to connect.

Sometimes conveying our hopes, dreams, or fears in a genuine way can take months. If we're willing to be the first mover and be vulnerable to others, we once again expose ourselves to learning emotionally. To build relationships which get better with time, we must be willing to continually display vulnerability. CEOs are vulnerable at the right times; they demonstrate what they'd like others to do through their own actions. They lead by example.

As we build this practice and extend vulnerability, we'll sometimes stumble on the true nature of other people. When this happens, we might discover we have relationships with people who don't want to change, are evil, or who want to take advantage of our vulnerability. At this point, we face either an uphill battle or we decide to move on from the relationship.

Ultimately, we will become the emotional average of the five people we spend the most time with. To keep a healthy average, sometimes we need to prune our relationships. We can continually extend the olive branch of vulnerability, but we should never be afraid to break off relationships with those who take us for granted.

An Uncomfortable Truth

The emotional average of the five people we spend the most time with sometimes doesn't hit home until we're around true CEOs or masters of their emotional states. If we insist on hanging around the best of the best, we may feel more insecure, but we'll be forced to improve our own emotional intelligence more quickly.

If we have faith in the CEO practice, we'll soon find ourselves leveling up, and discovering deeper emotional connections with our spouses, friends, family and colleagues. When we grow emotionally, we often discover those around us have been waiting for someone to give them permission to do the same.

10

The Satisfied Relationship...Impossible? Or Inevitable?

"Don't it always seem to go, that you don't know what you've got, 'til it's gone." –Joni Mitchell

"Happy!? Anybody can be happy. What's the purpose of that?" –Bob Dylan

An Aggressive Stoic

Around the time Christ was born in Ancient Rome, there was an entrepreneur turned statesman named Lucius Seneca. He was a philosopher who counted himself among the Roman Stoics. Seneca started from humble beginnings and rose to become one of the wealthiest and most powerful men in the world at the time. Some people say he was like a modern day investment banker, but he was a bit more difficult to classify than that. He was an entrepreneur who built up a fortune, lost most of it and faced exile for years, and eventually returned to Rome and rebuilt his fortune. Seneca was an adversity alchemist.

Towards the peak of his power, he was a bit like a venture capitalist or entrepreneur-in-residence for the Roman state. Seneca tutored the Roman Emperor, Nero, and was regarded by many as one of the wisest men in Rome.

Before Seneca's life ended, he sought to pass on much of his knowledge to one of his friends and pupils, Lucilius. Seneca's success and popularity had turned him into an enemy of the state, and towards the end of his life he was facing death for the crime of being *too* good. What the corrupt Emperor didn't count on was that Seneca was a master of long term thinking, so much so that he strategized as to how his teachings would survive and grow after he was gone. He knew the state would confiscate and destroy nearly everything he owned, so he needed a way to make his teachings antifragile. He compiled many of his teachings and philosophy into letters, which he strategically mailed to his friend, Lucilius. Lucilius was talented enough to ensure the letters survived, but not so talented that he had attracted undue attention from the state. Those letters which Seneca wrote to Lucilius became a book called *Letters from a Stoic,* or sometimes called *Seneca's Epistles.*

In letter IX of *Seneca's Epistles* he writes to Lucilius and relays with admiration the story of Stilbo, a practicing Roman Stoic. Being a practicing stoic meant that he would occasionally go through the mental practice of writing things off, or preparing in advance for the loss of certain things. We've seen earlier that reflecting on our own mortality is a never ending fresh water well for our motivational thirst. The practice of considering losing what...and who... we love is no different. This practice forces us into the present. It forces us into engagement and to properly value those we love today, while we still have them. The story Seneca relayed to Lucilius was by no means callous or flippant, but done with the greatest admiration. Stilbo had been away from home and when he returned, found barbarians had sacked his city. His wife and his entire family had been killed. This was the madness of those ancient times. In assessing the damaged city, Stilbo was asked, "*What have you lost*?" Stilbo answered, "*Nihil perditi. Omnia mea mecum sunt*!" Or, in English: "*I have lost nothing. My goods are all with me*."

At first glance it may certainly seem that Stilbo's response, and Seneca's praise of it, is incredibly cold.

But to understand the deeper meaning of Stilbo's response, we must investigate the Stoic philosophy towards life and relationships. We must recognize that the Stoics had an ideal mental state they were pursuing in life. That state was *apatheia*.

Apatheia is the mental state of being undisturbed by lesser emotions, a state where nothing can be taken from us because we have properly valued everything and everyone while they were with us. The Latin word "apatheia" doesn't relate to "apathy," but translates most closely to the English word "equanimity."

This is a mental state where we are not present day slaves to future losses. We don't let petty annoyances and passions direct our aims. We accept that losses will come, and distressing events will happen. It's only with that recognition that we have the willpower to place the proper value on every interaction with those we love while they're here. The Stoics recognized that the most prudent way to alleviate the sting of future losses lies in current interactions with the people we love.

Seneca praised the story of Stilbo in his letters to Lucilius as a poignant reminder. He also wanted to ensure that Stilbo's proud and courageous stance against fate would not go unrecorded by history.

The CEO practice is the first step in pursuing apatheia. It's the first step of recognizing our human condition and fighting to place the proper value on things as we go.

"...If a man can reduce his needs to zero, he is truly free: there is nothing that can be taken from him and nothing anyone can do to hurt him." –John Boyd

Anyone Can Be Happy

In popular culture, there has been a lot of talk about "happiness." Happiness is an easy sell. It's what some people would like us to believe is a noble or important goal. We're told that "happiness" is what everyone should be after. Of course, many people might have a definition of happiness which works well for them, and this is fine. Some people might be thinking of happiness as what Aristotle called the good life. But sadly, many people gravitate towards the word "happiness" as a form of socially acceptable mindlessness. They become tantalized by happiness, which becomes a willful ignorance of their own potential, a disregard for humanity as a whole, and a blind eye towards the horrors that life and fate hold.

Life contains suffering, loss, and death. The only control we have over those events is in the present moment. It's through the painful recognition of our own mortality, and that of those we love, that we become true empathetic stewards of time.

For humans, apatheia ultimately is not an attainable goal. It's impossible to always value everything appropriately in the present. It's such a lofty goal that it's not reachable in our human condition. That's the point. It often takes the sting of many losses until we discover and strive to reinforce true appreciation for what we have. When we harbor huge, massive, and sometimes impossible goals, we are lifted up to achieve far more than if we had been "reasonable" in our goal setting. Being reasonable or realistic rarely has a useful purpose.

The person with the courage to pursue the ideal state of apatheia is rewarded by progress. It's only by having audacious goals that we make any progress at all. It's in the pursuit of these things that we find a balanced satisfaction, and not the ephemeral notion of "happiness." The pursuit of the unattainable quietly forces us to make progress… every single day.

This place of continual struggle for progress is the only way to look back on the past and be satisfied. Pursuit of apatheia in the present means we create a store of positive memories which we get to hold onto and can reflect on with satisfaction for a life well lived. We only grow wiser with age, and earning satisfaction stores up golden memories for old age.

Weak people lacking courage are content in a state of happiness. Wise and courageous people know they will lose everything.

The person with the will to pursue apatheia is rewarded with the satisfaction of a life fully lived. No matter what blow fate brings, his relationships have been filled with life, and his *goods are always with him.*

There are two central practices useful in fighting our way towards apatheia. These practices are especially useful in relationships: curiosity and growth. These practices help us spark better connections, prompt more meaningful conversations, and prevent wasting time in arguments.

When Curiosity Blends with Growth

Curiosity doesn't mean just mindlessly asking questions, but asking well thought out questions. Asking great questions comes from writing lists of possible questions, practicing them beforehand, and then picking the best ones to help us get to know our partner better.

What stops us from being curious is often our ego. A certain amount of ego dissolution is necessary to begin the process. Once we sidestep ego, we can remove the fear which accompanies asking challenging questions—the fear of being wrong, of saying something stupid, of misspeaking. Genuine curiosity sparks meaningful connections in relationships.

Consider this test: When was the last time someone was actively curious about our endeavors, preferences, likes, dislikes, aspirations or ambitions? These types of personal, curious questions are rarely ever asked. When they are presented to us, we remember them forever.

These are the interactions our spouse will remember for years. Curiosity precedes the type of connections and bonds which will endure whatever fate throws our way. Curiosity is the key to empathy, which is the cornerstone on which everything else is built.

When we stay curious, we multiply the moments and chances we have to connect, reconnect, and build our relationship. Our minds are naturally creative in this state. Curiosity breaks the barriers of imagined differences, and we see most of our problems as fleeting.

Curiosity keeps us resourceful. This practice has the by-product of keeping us open to opportunity in life. The curious person is more likely to stumble on secrets, compelling anecdotes, or find a reason to keep going when all looks hopeless.

Now we'll mix growth and curiosity to help create satisfaction. Here are three equations to consider regarding how curiosity and growth work together to help move us towards apatheia:

An Equation for Apatheia

Curiosity Without Growth = Happiness, but Eventually Pain. This inevitably brings the regret of looking back and wishing we'd done more.

Growth = Satisfaction and a Past which is Bearable to Examine. Humans are only satisfied, or blissful, when we make progress.

Curiosity + Growth = Satisfaction. This is one of the master equations for relationships. It leads towards uplifting relationships which get better with time, and offers us movement towards apatheia.

Growth or progress in a relationship arrives when we take an enlightened view of our own mortality. When we set the goal of apatheia, we're inspired to make progress. But if growth and curiosity are key elements of successful relationships, why don't more people choose them?

The answers are all over the place. My guess is that it's because it feels unnatural to do at first. It's difficult to put these practices into action. It requires huge amounts of faith that the upfront investment will prevent future pain. Undertaking a path of growth and curiosity together requires enduring periods of uncertainty.

The person who remains ignorant of the benefits of these practices usually underestimates the future suffering from not engaging in them today.

When we seek to grow, we'll naturally feel uncertainty and ignorance while we're moving out of our element. Many people think this is a signal they're stupid or not cut out for it. This is the point where they typically shut down, become defensive, attack others, or give up the pursuit. This person continues to remain ignorant, or just settles for a happy ignorance.

It's easy to think of our intelligence or abilities as fixed. It's natural to fear that if we try to learn more things, we'll overwhelm ourselves. The only way to grow is through continually stretching our abilities and capabilities in the face of uncertainty.

The person who grows understands that feeling ignorant about something is just a precursor for growth. It's only through enduring temporary feelings of ignorance that we're able to learn anything.

After a short time on this path, pieces of knowledge gained in uncertainty have a tendency to compound. We're never afraid to ask the questions that matter. Soon, we connect deeply in relationships.

Growing personally is the best way we can show how committed we are to our relationships. The best way we can thank a spouse or friend who has supported us is by growing and becoming curious. When both parties in a relationship are aiming towards satisfaction or apatheia, the load of the work is lightened.

Strategy and Action

"What's the world's greatest lie?", the boy asked completely surprised. "It's this (said the mysterious king), that at a certain point in our lives, we lose control of what's happening to us, and our lives become governed by fate. That is the world's greatest lie." –Paul Coelho, The Alchemist

"There is nothing on this earth more to be prized than true friendship." –Thomas Aquinas

The Relationships That Matter

One of the godfather's of logic and humanism was Aristotle. Two of the quotes attributed to him perfectly articulate the concept of apatheia and satisfaction, only in relationships. These quotes are both relevant for deep friendship, as well as for deep romantic relationships:

"Friendship is a single soul dwelling in two bodies."
"Love is composed of a single soul inhabiting two bodies."

The lists, questions, and prompts below can be completed with or without our significant other. We can even complete them by writing our answers out on paper and leaving them for our spouse to find. How would we feel about losing what we love? This is the Stoic practice of mentally preparing for loss today, in order to care and value it properly today.

How does it feel?

1.
2.
3.

After hypothesizing about feelings from the future loss, what can we do today to better value our relationships? How can we effectively give the middle finger to fate, and all that we cannot control?

1.
2.
3.

Now brainstorm 5 questions we can ask which demonstrate genuine curiosity.

We can try to work them naturally into conversation. These can be questions we don't know the answers to, or have been meaning to ask for years.

1.
2.
3.
4.
5.

Reflecting on Growth

A great quarterly exercise to help build satisfaction in our relationships is to list how we've grown together.

What are three things that, as a couple, we've achieved or have accomplished together?

1.
2.
3.

Remember: Everything we love will eventually be taken from us.

Curiosity + Growth = Satisfaction (or movement towards apatheia and a middle finger to everything outside our control). Growth and curiosity aren't always easy, and in the next section we'll accelerate the growth through physical momentum.

11

Finding, Keeping, and Building The Relationships That Matter

"Adversity is like carpet bombing false relationships...When you face significant adversity...There is a great sunlight that causes indifferent or false friends to scurry in the darkness away from you." –Stefan Molyneux

"A body in motion will stay in motion, while a body at rest will remain at rest." –Sir Isaac Newton

More Room for Activities!

In the movie *Step Brothers*, Will Ferrell and John Reily's characters discover the beds in their room have some serious potential. If they stacked them...dear God...they would fit together perfectly and become bunk beds! This would leave Will and John with ample room for every kind of "activity" imaginable!

The scene is a funny reminder that exciting moments in relationships only come after we're in motion. Only once we're focused on activities, getting out, and trying new things, do we give ourselves exposure to serendipity. It's only through faith and trust in this process that we build the bonds and memories which help relationships thrive. Storing up these memories is like hibernating for winter and all the relationship challenges we will inevitably face.

When we become childlike by making space or time for activities, we can't help but create great memories. It's these moments which help us properly value our relationships while we grow.

Movement leads to better moments, and better moments help expose us to serendipity. In the above story, serendipity is discovering that the beds stack together PERFECTLY to make bunk beds.

In our own relationships, serendipity might arrive while we're on a road trip, taking a class together, on a double date, in a yoga class together, or discovering a business idea which our spouse or friends are perfectly suited to pull off. What is the equation here?

Movement + Exposure = Moments and Serendipity

To get moving we need a willingness to put ourselves out there and not be afraid to ask our parents if we are allowed to stack our beds on top of one another. It's a willingness to open up and be ourselves, which makes us appear "lucky" in relationships. We can't manufacture serendipity and joy on demand, but we can increase our chances of stumbling upon them. If we look back on all the best moments in our relationships, we'll likely see that many came without meticulous planning.

This should take the pressure off us. By removing pressure, we free up energy we can use to focus only on what we can control-precursors to serendipity. In this case, the precursors for the great things we can't plan are movement and activities.

Working on becoming a CEO who can connect strategically with people through curiosity and growth is great. But it's only when we get moving and increase the amount of activities we do, that we start increasing our chances of creating amazing experiences and memories in our relationships.

People often underestimate how movement is the single driver to break out of relationship challenges or stagnation. Here are two quick stories about faith in movement and activities from two entrepreneurs: Tony Hsieh and Chris Sacca.

First, let's take Hsieh's story. He has started many companies, including LinkExchange (sold to Microsoft), and is most famous for building and selling Zappos to Amazon. Hsieh is in the midst of building a technology hub in Las Vegas around the Zappos headquarters. Many of the buildings and outside areas are strategically designed to spark connections and activities, and to "encourage serendipity." The inspiration for manufacturing serendipity came when Hsieh was reflecting back on his life. He discovered that many of his most pivotal moments and meetings came without planning. He had given himself exposure to serendipity through activities and chance meetings.[25]

The second story comes from Chris Sacca, who at the age of 25 was in debt for $2.1 million, in his own name.[26] How he got there is an entirely different story, but the amazing part was that he did not declare bankruptcy. He fought his way back and repaid the debt. He did everything from voiceover gigs on Elance to sneaking into networking events through the kitchen because he had no money to pay the fee at the door. From there, he continued to chip away at his debt. He leveled up in his career as he went, becoming an executive at Speedera Networks (acquired by Akamai Technologies) and went on to become the Head of Special Initiatives at Google. Now Sacca is the principal of the venture capital firm, Lowercase Capital, which has invested in dozens of companies and owns large stakes in Twitter and Uber.[27] Going from negative $2.1 million to over $1 billion requires a huge amount of sheer faith and determination. It could not have been done without movement and exposure to upside.

One of my favorite quotes from Sacca is in regard to the CEOs who actually innovate. So often we conceptualize CEOs as people who are hard-chargers. We mistakenly think it is the 100 hour work weeks which produce the innovations we laud.

In reality, these types usually burn out early. They destroy their bodies and minds, and poison their relationships. Sacca affirms this with a great quote on the type of entrepreneurs he looks to invest in, "…it's the people who know to turn off the switch and go outside…it makes for well rounded people and ideas come from those people… it's people who have balanced relationships that come up with the cool shit."[28]

So let's give ourselves exposure to the amazing things by getting out and getting in motion. Movement is one of the most powerful fuels we can add to our lives and relationships. It's when we're out paddle-boarding with our spouse that we'll create a great moment. It's when we're going skydiving with a friend that we'll create a memory we'll have forever.

It's not the 14-hour days of work that build amazing relationships. It's not through staying at home watching the same television show everyone else is watching. If that's all we do, we'll end up looking back in frustration at what we let pass us by.

When we undertake activities and new experiences that add positive memories and endorphins into our relationships, the quality of our relationship skyrockets. By keeping this balance, we effectively give our relationships the best chance to thrive over the long term.

So let's jump into the strategy and action section which has more ideas for movement, activities, and exposure to serendipity.

Strategy and Action

"The man who moves a mountain begins by carrying away small stones." –Confucius

"It's not a failure if you have to resort to Plan B. Or even Plan Z. You will be stronger, happier, the people around you will benefit, and society will benefit from having toughened you up for what you now deliver to it. It's the BEGINNING of success when you start having multiple plans." –James Altucher

From Fridge to Friends

The best moments are never planned. They're the result of getting moving without too much thinking, and giving ourselves exposure to luck.

I was thinking of calling this strategy and action section *Awesome Activities That Accelerate and Engineer Moments of Serendipity*, but that's just ridiculous. A much better label for the lists that appear below is "*Fridge Lists.*"

This highly technical term refers to the fact that these lists are best utilized when they're literally written on paper and stuck to the refrigerator. They stare us in the face several times a day and remind us to stay in motion. Innovators get out and do; they connect and are constantly building relationships through activities which pump out endorphins.

Some of the things my wife and I have on our Fridge Lists of activities include categories: Events, Conferences, Meet-ups, Physical

Adventures, and Travel Ideas.

Are many of the ideas listed below simple? You bet. But are they simple to do? No, of course not. Otherwise, everyone would have amazing, interesting relationships.

Getting in motion is the hard part. It's easy once we're going. It can be challenging, but we'll be happier when we're in motion. Below are some ideas on how to build your own *Fridge Lists*.

Events, Conferences, and Meet-ups:

- What conferences are coming up which we're interested in? Many of the big technology companies you might want to work for have conferences and they're a blast. Go to them.
- Meet-ups www.meetups.com
- Speaking Events and Book Signings… What ones are coming up which we're excited about?
- Comedy Tours (your local spot here)
- Plays. Yes, plays. *The Book of Mormon* on Broadway will make you laugh for two hours straight. Do that with someone you love and your relationship is immediately better.
- Co-Founders of (wherever you're located) CoFoundersLab.com Hint: Stay away from the people on the site looking to get rich. Find people to make friends with whom you can trust while you're building a company.
- Veterans' Meet-ups at your college or local city—if there isn't one nearby, create your own.
- Concerts. We start with a list of the ones each of us want to attend, and then narrow our way down to the most mutually agreeable ones. I've found the best way to ensure we actually go is to buy tickets on the spot.

A subsection of this list is our "Learn to" list:

This is a list of things we want to learn…

- Paint
- Hunt
- Make Wine
- Make Beer
- Roast Coffee

Maybe for you it's learning a language, a musical instrument or a craft, or how to garden, fix cars, or do home repairs. Whatever it is, learning something new stimulates the brain and widens social networks.

Physical Adventures:

- Take the 3DR Drone out and film and practice flying in an open area.
- Record with a GoPro and tinker.
- Local Gym Classes: Cycle, Battle Ropes, etc.
- Hikes (We list specific ones in our area and new ones we want to go on)
- Paddle Boarding
- Kite Surfing
- Yoga
- Dance Class (Massive, massive points scored with the spouse for this one)
- Trampoline Indoor Arena
- Tandem Skydiving

Travel Ideas:

- These can be longer day trips, mini-vacations, or even month- or year-long sabbaticals.
- Cross-country drive trip to celebrate being completely location independent
- Snowboarding Trip (Airbnb rooms for cheaper travel)
- State Parks
- Highly-Rated Trip Advisor places we want to go
- Winery / Breweries we want to see along the way
- Snowboarding group trip (group trips are key to saving tons of money)
- Europe trip next year!

The point of these lists is to make the friction required to get moving easier. Most people willingly allow themselves to be bombarded by advertisements all day long. Why not change it up and advertise to ourselves? Fridge Lists help direct and target advertisements to ourselves to create outcomes we want. Lists all over the fridge are great places to plaster advertisements, reminders, and ideas to reduce the friction required to get moving.

Also, when we're in the process of finding new friends or building new relationships, having fun and interesting events already listed and ready to go makes connecting easier. These types of activities and movement are much healthier than just the usual happy hours. There are fewer barriers to connecting in a meaningful way if we can quickly glance at these lists on our fridge and shoot a text or email that says:

"Hey, (event, meet-up, activity) is coming up on the (date)—you want to go?"

The *Fridge Lists* are the best way to manufacture serendipity, speed up the rate at which life gets interesting, and make great connections.

THE CHALLENGE

The challenge is forcing ourselves into motion even when we don't feel like it. What's so hard about this practice is we have to force ourselves to get started. Whether it's dangling a carrot in front of us or having our spouse schedule things because we know we never will, we have to ensure these activities get scheduled and we get in motion.

The more we can build our relationships through shared experiences, the less likely future experiences will break us apart. This also helps create the type of relationships where taking risks becomes familiar. We'll never be at a place where we innovate, start a business, or do great work unless we practice taking series of adventures and small risks. Whether it's starting a business, having a business fail, climbing out of debt, or rebuilding after a deployment, all these trials are easier with lots of fun memories to look back on. And if we have taken risks and succeeded with our spouses, we are much more tolerant to stress when trouble comes.

So now that we have plenty of ideas in regards to becoming a CEO who builds great relationships, let's jump to the specific skills the world pays a premium for and the future will demand.

Part 4: Skills

12

Global Skills That Will Always Pay The Bills

"In today's global economy here is what is scarce... Quality labor with unique skills... Here is what is not secure these days: Unskilled labor, as more countries join the global economy..."
–Tyler Cowen

"...Capitalism and competition are opposites...under perfect competition all profits get competed away. The lesson for entrepreneurs is clear: if you want to create and capture lasting value, don't build an undifferentiated commodity business." – Peter Thiel

Waiting for a Flight

It was after a conference, and I was at an airport. I had time to kill before the flight so I went to grab a coffee. At the cafe I ran into another veteran I had met at the event, and we stopped to talk. He was getting ready to graduate college with a bachelor's degree in business administration, and was nervous about entering a rough job market.

We chatted for a while, and he shot off business and job related questions like lightning. He had debt from his school days and was nervous because he hated his current internship. He wanted to get a job doing something— anything that was interesting.

He seemed to be a smart guy. I asked him if he had thick skin. He laughed and said he did. I brought up the fact that around 368,000 other American students were currently majoring in business and getting ready to graduate. He was in a field with intense competition. We talked about the business program he was attending, and I was shocked the more I learned about the curriculum. Calling it outdated was an understatement. He was getting ready to graduate with no portfolio of work, no consulting gigs, no skills with basic business software like excel, and a network of about 20 connections on LinkedIn. On top of all of this, he was in massive debt.

I offered ideas such as attending a developer boot camp, or mastering a brand new software program like Tableau, doing consulting work and leveraging that for a reference, or creating a portfolio and body of work which would bring in interest from headhunters. We strategized for a bit and then parted our ways for our flights. I haven't heard from him since then, and I hope he's doing well now.

I can't help but feel frustrated thinking about his plight. He trusted that the four years he spent in school were leading somewhere worthwhile. He trusted that what he was learning would be relevant when he graduated. Now, more than ever, the heuristic "trust but verify" applies to any traditional education program.

I wonder if this trend will continue. Students are devoting on average 5.6 years of their lives to obtain four-year degrees, only to leave college and immediately need retraining.

The industrial age has been a massive engine of progress for the United States, but it baffles me why our education model hasn't progressed. I think we're probably just in the last stages of thinking college is the best path for learning skills.

Unskilled Labor = Commodity

One of the last myths of this age is that there are magic degrees which guarantee high-paying jobs. Sadly, there is no such insurance policy anymore, and those who collectively owe over $1.3 trillion in student loans can attest to this.

In many ways, the skills part of this book is a pathway through the mess of higher education. These chapters are about engineering our way into high-demand fields. It's from these fields that we'll quite literally have a chance to rebuild America.

We're still in the early days of technology and the Internet, but the cultural shift they have brought is revolutionary and will irrevocably change the world. Let's take a more in depth look at what exactly is going on in the world.

The first quote at the beginning of this chapter comes from Economist Tyler Cowen, who wrote a book called *Average is Over*. To sum up the entire book in a sentence: The old days, when "average" could get you an "above average" job, are over. These days, seeking old traditions or average leads to a pit of quicksand of marginalization and commoditization.

Unskilled or minimally-skilled labor has never been cheaper and more abundant. Machines, software, and technology will continue to consume white collar jobs. We are entering an area of intense competition for "middle class jobs." These jobs have been, for the last fifty years, accessible only with a bachelor's or master's degree. While many of these jobs might be accessible with a bachelors or masters degree, there are many accelerated paths outside of them.

We're living now in an economy where employers ask, "How can you help solve my problems right now?" These employers, including Google, Khan Academy, and Facebook, don't care about degrees or GPAs. They want people with a baseline IQ, high emotional intelligence, creativity, leadership ability, and SKILLS.

Those who want to rebuild America must be willing to aggressively learn new skills and showcase their expertise via a portfolio or body of work. The good news for veterans is we've built some skill sets that are incredibly valuable, such as the ability to tolerate discomfort and uncertainty. We know when to speak up and when to shut up. Veterans have learned about leadership under the tutorship, stakes, and incentives of life and death. These teachers completely trump any classroom stakes and incentives where the greatest stakes and incentives might be fear of a bad grade. The leadership lessons learned through facing life and death decisions are ever-present guides and accelerants for acquiring skills. In America, where many people are afflicted with many diseases of comfort, Veterans must remember that we have foundational skills no one else has developed.

We want to leverage these abilities by combining them with the immediate skills employers are ready to pay for right now.

The second quote at the beginning of this chapter comes from entrepreneur and venture capitalist Peter Thiel. In his groundbreaking book, *Zero to One,* he introduces many important messages, namely, the importance of creating technology which take us from 0 to 1. The definition of technology is something that lets us do more with less. It takes us from where we were to a new place in a new way where we can accomplish more with fewer resources.

Everything outside of technology in a globalized world takes us from 1 to n. This means many businesses take existing technologies and just spread them all over the world. This doesn't appear to be a bad thing at first, but after further examination, it isn't sustainable. If we merely want to take all the existing technology we have created and spread it all over the rest of the world (1 to n), we'll destroy the planet. Our existing standard of living in the West cannot be extended everywhere using current technology.

If every Indian and Chinese person starts consuming the same amount of energy the average American does, mining the sheer amount of coal necessary will be impossible. The pollution from burning that much coal would decimate their countries.

So if average is over, and simply spreading the technologies we currently have over the rest of the world is deadly, what are we supposed to do? Joining and aligning with the companies here in America which are seeking to create innovations to take us from 0 to 1 is crucial. These companies are literally rebuilding America and creating a better future for humanity in the process. But we can't join these revolutions without bringing serious skills to the table.

The survival of America, and ultimately humanity, depends on those brave enough to join up with industries, companies, and groups of people pushing to go from 0 to 1. It's no accident that the huge majority of these companies, industries, and groups of people creating and building new technologies are American. These groups are taking small amounts of capital and turning them into huge value that adds tremendously to our country's GDP.

The tactical, individualized answer to avoiding the commodity trap of average, and to building the skills necessary to rebuild America (and save the world), is through layering skills.

It's the practice of starting with the skills offering us the greatest upside, and which are in highest demand. From there, we strategically layer on more skills until we become a monopoly of one. Along the way, we'll make ourselves vital to the industries going from 0 to 1, or be able to create our own technology which takes the world from 0 to 1. Now let's dive into the nitty gritty of how to do this.

The skills section of this book is designed to be a buffet, so feel free to choose whatever section is most applicable to you right now, or jump around as you see fit. The menu:

Chapter 13: The Core Skills
Chapter 14: Mastery of Technology
Chapter 15: Reverse Engineering Rebuilding America
Chapter 16: College or Not?
Chapter 17: Return of the Polymath, Layering Skills

We'll see that this progression is simple and available to everyone. The problem (or opportunity, depending on your point of view) is most people aren't patient enough to layer skills in a strategic, economical fashion. They're tempted by the allure of instant cultural approval if they're only working towards their "passion." Or temptations of making millions of dollars from their ideas alone steer them away from the hard work of strategically acquiring marketable skills. Being an idea machine is a great thing. Being an idea machine which systematically acquires the skills necessary to turn ideas into their physical manifestations is even better. Layering skills is about generating ideas and then becoming expert at the execution required to sell, or alchemize them into reality.

Those who will succeed today and in the future will be those determined enough to build up the core skills, master technology, layer skills, find the best path to learn, and place themselves in positions in industries creating technology. It's this type of person which has a chance of actually finding their true passion. It's this type of person who is patient enough, who naturally becomes more entrepreneurial along the journey. They are the ones who have a chance to begin rebuilding America. These entrepreneurial leaders will win a chance to work on the projects and technology which take the world from 0 to 1. Eventually, they will find themselves shoulder to shoulder with the right relationships and virtuous groups of people. It's from these springboards that they will be able to start their own businesses or participate in creating the technologies which will take humanity from 0 to 1.

Strategy and Action

"...pointlessly dicking around is what most people have confused for work..." –Ryan Holiday

"Everybody wants you to be boring, so they don't have to be interesting." –Stefan Molyneux

Skills and Competency Triggers

This strategy and action section is a brief overview to get us acquainted with the language, terminology, and mindset of how to go about acquiring skills. The goal is to acquire them in such a way that we become a monopoly of one. This pursuit of a monopoly of one means gaining self knowledge and becoming so skilled that we become able to add tremendous value to the lives of those we care about most.

It's this type of path and skill acquisition which makes us antifragile. It's this pursuit which gave our founding fathers the ability to start America. They were true self taught polymaths. After they obsessively acquired enough skills and resources, they eventually launched the biggest startup experiment the world has ever seen—America. The path of self-directed learning and skill acquisition is what built our country, and it's how we can start rebuilding it today. This is an overview of the chapter buffet contained in the skills section of the book:

The Core Skills (Competency Triggers):

As we've seen, the foundation of our skills begins with our personal health and the CEO practice.

The next addition to our set of core skills is the most crucial. These skills are sometimes called competency triggers. Competency triggers include skills at reading (a lot), ability to write well (a few hundred words each day), and both of these practices fuel the third competency trigger, which is the ability to speak more articulately.

When we execute or showcase these competency triggers to other people with high levels of competency, we trigger a type of recognition which acts as an instant credential. This recognition can save us from having to explain much of anything. Competency triggers are like resumes of everything we've ever done that can prove and suggest the type of work we're capable of almost instantaneously. The higher our ability and skill at these competency triggers, the less we have to sell or convince. The leader who is continually signaling competence and mastery doesn't have to expend energy selling their ideas. Competency triggers help build high quality ideas, market them, sell them, and attract people with other high levels of competency.

For instance:

Reading great material signals that we enjoy learning, can teach ourselves things, and likely have interesting things to talk about. Entrepreneurial companies and businesses want to associate with, hire, and offer opportunities to people who are always learning (reading the right stuff).

Writing is another practice to build and signal competency. A vast amount of communication is done via email and chat correspondence online. An email which shows our competency in a few lines can completely change our lives. Leveling up as writers is a must. Plus, it's a way to clarify our thinking and helps build the next competency trigger, which is speaking.

Speaking includes being able to articulate ideas in a contextual fashion. This includes the ability to speak in front of groups, peers, or pitch a solution to a boss or customer. The ability to speak well is developed by lots of reading and writing. Speaking well also builds the ability to listen well. Speaking and listening synthesize to help us make deals, rouse an audience, and persuade others of their true potential. *Language Intelligence: Lessons on Persuasion from Jesus, Shakespeare, Lincoln and Lady Gaga* by Joe Romm is a useful book on this subject.

Competency triggers are often better substitutes than any formal credential. If we work on perfecting these, they'll be recognized by the right people at the right times.

Uber-successful people have mastered competency triggers. These successful people filter the people they choose to do business with, interact with, or hire, by nature of who displays expertise via competency triggers.

Mastery of Technology

The ability to use the right technology to learn and create is crucial. We want to come up with rules for adopting and using technology to ensure it serves us. As a culture and society we haven't really applied any rules to how we use and adopt technology. It's important to create individualized guidelines for ourselves, so we're not used by technology. We want to develop ways to use technology strategically, instead of mindlessly being controlled by it. Think of all the requests you get via Facebook to play games with $99.99 in app purchases of virtual currency.

As more technology is introduced into our lives via smart objects, sensors, and heads up displays, it's never been more important to develop criteria for use and adoption. We will look to the Amish for help with this one, as Jebediah has much to teach us.

Marketable, Tangible Skills

These are skills we develop which have a market value. These are things we learn which boost our earning power and increase our ability to transform our time into values for others. We know we're learning marketable skills when people begin asking to pay us for them.

We want to learn skills with the highest demand in the marketplace which also correspond to our interests. When we follow this path, we are acting in a virtuous manner because the increased demand signals a shortage of those pursuing the skill set. This small step benefits all the other people in our economy. We'll discover how we can quantify and advertise our skills to create opportunity generators online. By adding these skills to our online presence (social networks, website, content we create), we'll get offers to be paid for our skills by places in need of our services. Or, we'll get information about what skills we need to learn in order to be paid or command a premium for our abilities.

College or Not?

This chapter is about deciphering the right training needed for the right skills. The old model of college is largely broken, and higher education is now not the only way to build marketable skills. We need to rigorously test what the best route is for us when it comes to thinking about college. We do this in Chapter 16, titled "College or Not."

Reverse Engineering Rebuilding America

This chapter explores how to find the companies rebuilding America, and how to take advantage of opportunities at these companies. This is about aligning our interests with organizations creating 0 to 1 technology, or breaking into companies which are contributing to our GDP and creating jobs (hence rebuilding and solidifying American power). This chapter shows how we can put our effort into places which will allow us to contribute in the highest yield way possible. If some or all of the 2.8 million post- 9-11 veterans decide to align themselves and join these industries, the American economy will BOOM.

Layering Skills, Becoming Irreplaceable, and Evolving into Monopolies of One

The mindset of layering skills helps us build monopolies of one; or we could call this just finding ourselves, self-knowledge, or finding our true passion. The idea is to capitalize on our uniqueness again and again and again, until we've built up such an eclectic and specialized set of skills no one can compete with us. By becoming a monopoly of one, we give ourselves the ability to not only live incredibly rewarding lives, but also build the proficiency to eventually build and form our own companies, products and services, or deliver existing services in a 0 to 1 fashion.

This is how entrepreneurship is built. Layering tangible skills and becoming polymaths is the same way our country was founded, and it's how we will rebuild it.

13

The Core Skills For The Future

"I read every book and magazine I could. Heck, three bucks for a magazine, twenty bucks for a book. One good idea would lead to a customer or a solution, and those magazines and books paid for themselves many times over…In doing all the reading I learned a valuable lesson. Everything I read was public. Anyone could buy the same books and magazines. The same information was available to anyone who wanted it. Turns out most people didn't want it…" –Mark Cuban

"Read 500 pages like this every day. That's how knowledge works. It builds up, like compound interest. All of you can do it, but I guarantee not many of you will."
–Warren Buffet

Ask for Answers from Those You'll Believe

When I was in school I was frustrated because I only wanted to learn things which would help me in the real world. I only wanted to learn things I could apply to life. I wanted a set of skills which would unlock opportunities. I wondered what or how could I learn that which would serve me for life.

Eventually, I became frustrated enough that I decided to seek answers from people I wanted to emulate. I came up with a simple plan. I would find people online who had created or become something worthy of emulation. I'd study the hell out of them to find how they learned the habits and skills which led to the core of their success. What were the similarities or methods of learning which all these people cited as having most influenced their success?

Every time I stumbled onto somebody interesting or inspiring, I either asked him or her this question or searched for the answer online. Sometimes when I identified someone online, I scoured through everything they'd written or spoke about. In many cases, I found answers directly in their quotes, work, or things they had said in speeches or commencements. If I didn't see the answer, I simply reached out via email or phone, or met them in person and asked them what the most crucial personal skills they'd developed were.

This practice involved writing a countless number of emails to the people I was studying. It led to learning how to properly form questions, and seek to add value in ways which demonstrated competency triggers and warranted a response. This practice also led to meeting, conversing, and connecting with many of these high achievers in person, including a former President of the United States, a billionaire venture capitalist, CEOs of publicly traded companies, television show hosts, founders of businesses creating amazing technologies, venture capitalists, a multi-millionaire author, and many others. When I studied each of their back stories and the skills they had learned to get where they were at, the answers I found were amazingly uniform.

The responses and answers I found from these successful leaders, founders, entrepreneurs and executives could be boiled down into three skills they each had built:

- Learn to love reading; read a lot, and read a lot of the right stuff

- Practice making your writing better daily, as it will improve your thinking and speaking
- Improve your abilities to speak and listen, and be able to speak well in public

At first I was frustrated because many of their responses and paths seemed simplistic. But after more thinking, practicing, and doing, I could see the truth in what they were saying. I had wanted to find the fountainhead of skill development. The three core skills of reading, writing, and speaking were it. When I looked around me, it was as if this secret was highly visible to some, yet elusive to most who knew it, and refused to put it into practice.

The reasons why developing these skills are undervalued is likely because from an early age we're forced to do or practice them. Our mindsets towards reading and writing become tainted early on because we're asked to focus on subjects without any real world significance. We're usually given horribly boring things to read, followed by the assignment of writing about things we didn't want to read in the first place. Then that coercive experience is fully ingrained as unpleasant when we're asked to speak publicly in front of a room of our peers about what we didn't want to read, write, or say in the first place. It's no wonder we learn to undervalue these core skills.

Reading

"In my whole life, I have known no wise people who didn't read all the time—none, zero. You'd be amazed at how much Warren reads—at how much I read. My children laugh at me. They think I'm a book with a couple of legs sticking out." –Charlie Munger

"What I was given to study in school I have forgotten; what I decided to read on my own, I still remember." –Nassim Taleb

When reading is done right, it's pleasurable, it's entertaining, and it usually teaches a lesson. We receive and remember information when we choose what to read on our own volition. We can't enjoy reading when we're assigned to read by someone we don't want to be like. When we use our own common sense, and seek reading recommendations from those we respect, we start to discover the intellectual power of the practice.

Freedom to read what we want is the fountainhead of all other learning and achievements. Reading informs our thoughts, allows us to dive into the consciousness of great thinkers, and pays rewards of all proportion to the small upfront investment of a book. Reading distills our ideas into concise and crisp communication. It allows ideas to percolate, then bubble up and combine into new solutions to problems we face. It leads us to creative ways to solve every problem around us. It inspires and informs our writing. And, one of the lesser-known benefits of reading is that it improves our ability to speak. This is what all massively successful, intelligent, and creative people know. Reading is one of the competency triggers which are easy and cheap to build. But building this trigger in a society with incessant "entertainment" can be incredibly challenging. The person who becomes skilled and committed to reading in our age can completely spring ahead of the crowd who choose entertainment over erudition. Reading is the key to navigating the path towards a monopoly of one.

By reading more, we become entrepreneurial and no longer need education; we're able to find what comes after, which is learning. Through this simple habit, we will always be able to create options in our own life. We get the feeling of what works and what we remember, and are inspired by the intellectual curiosity of others. Reading helps us build our own capacity to focus. In a world where many bow their heads towards a Smartphone, the person who maintains the patience to read long form books is a rarity.

More people than ever do most of their reading online. Whatever random content gets thrown at them, they consume, and so they fill up on what everyone else is consuming. Consuming, of course, isn't learning. The person who decides to read books which have stood the test of time takes advantage of the ultimate filtering mechanism. When people pay money for books they automatically gain stakes and incentives to learn. When they consume content online, they are incentivized usually by advertising dollars and impressions, which degrade our willpower.

Reading is the one practice guaranteed to relax our minds and take us away from anxieties and the noise of other media. Any media which is advertising based—TV, online news, radio, etc. — is designed to make us agitated. What people call "news" usually doesn't benefit us at all. It's most often just the entire noise of the world funneled into one place. It's packaged as news, and we're taught in school that it's important to be "well informed about current events." Our human evolution makes us good at learning and dealing with local events, but when we try to make sense of the daily news from around the entire world, we predispose ourselves to becoming victims of fear, uncertainty, and doubt. Reading is that rare advertising-free medium which filters and distills what actually matters out of the daily noise of the news.

The primary reason so many of us have short attention spans is because we don't read books as much as we used to. We're becoming used to game-like interactions and the endless stream of mindless entertainment available online. There is nothing wrong with some of that, but over time it corrupts our attention span. After a certain level of saturation, few people ever recover.

The nickname I love for physical books is the "perfect technology." No matter what happens, the battery life is always good to go. Books cost money so they come with incentives and stakes to read them. Plus, they take up space, so if we don't read them we're forced to suffer the consequences of looking at unread books we've paid money for (more incentives to read them). Books are now more affordable and of higher quality and greater selection than at any other time in the history of the world. Unlike tablets or reading devices, physical books help force us to only do one thing at a time. They prevent us from dealing with the urge of tapping away from the book to the Internet, an app, or a game. Reading, in practice, is a bit like practicing meditation or mindfulness. When we practice being fully immersed in reading, we're meditating or practicing mindfulness. A physical book is, indeed, the perfect technology.

When someone who reads is presented with an opportunity, they are able to draw on the wisdom of hundreds of experts and thousands of years. When opportunities present themselves to someone who doesn't read, they only have a narrow slice of experience or current events to draw from. Wise people love to talk about what they're learning, and oftentimes this means what they're reading. By preparing and reading now, we'll have something to say to the person we never thought we'd meet. We have a competency trigger to signal appropriately to the right people, at the right time.

What looks like luck to the causal observer is simply being prepared for opportunity. There isn't a cheaper or more effective way to capitalize on opportunities which present themselves than by becoming well read.

Writing

What I learned found its way on the paper, so I could dictate it, articulate it... –Nasir Jones

Good writers have two things in common: they prefer to be understood rather than admired; and they do not write for knowing and over-acute readers. –Friedrich Nietzsche

Writing well is the difference between sparking an interaction with a person that could change our lives, and missing that opportunity. With so much of our correspondence done online and through email, improving our skill at writing can pay dividends.

In business, consider the act of making a single sale or landing a meeting with a decision maker. If you write an email that gets this person's attention, sparks their interest, displays your competence and clearly explains what solution you have, you're literally transforming words into monetary value.

Writing well adds clarity to our thought processes. It builds the ability to articulate ourselves at pivotal moments during negotiations. Virtually all of humanities problems, wars, and generalized dysfunction stem from an inability to negotiate. Writing enhances our ability to become an interpreter between parties and bring good reason where it is needed most. By becoming a better writer, we also set the stage to improve our speaking.

Practicing writing before reading is almost impossible. Writing is something which begins to flow naturally out of us after a sufficient amount of reading. So we needn't worry about practicing our writing until we've mastered the habit of reading. But there are small areas where we can begin to practice writing which can improve our lives. Consider the first: writing better emails.

Researchers have studied students writing for assignments at school, versus writing emails to friends. They've found that students writing emails to friends write more clearly and succinctly than for any assignment. This isn't much of a surprise considering what we know about stakes and incentives. This also suggests a great place to begin the habit of improving our own writing. In the strategy and action section we cover a formula many great writers and advertisers use to keep their writing enjoyable, informative, and valuable to the reader.

The Gateway

Maybe you think you'll always hate reading or writing. I did too, until I had the proper stakes and incentives. I remember all the negative conditioning I'd gone through when I first learned to read and write. I had to read things I didn't want to, then write about them and be critiqued for grammar and given a grade. All about something I didn't want to read in the first place! It's no wonder most people hate reading and writing by the time they leave public school.

I started the reading and writing practice during my last year in the military. I wish I had started it sooner, but was glad I got started at all. At first, I set out to read a book a week, and probably averaged writing only 200+ words a day. This was really small, but easy enough that I could accomplish it. The benefit of a small goal like this is it doesn't feel daunting to start. And it's starting which matters most.

In a short time, the practice yielded clarity of thought and boosts of creativity. I learned more in a single year than I did in all 15.5 years of public education (including college). I can credit improved reading and writing skills with job offers, connections with amazing people, starting my own businesses, and getting married to a wife I love. My self-awareness jumped, my anxieties decreased, and I was able to start enjoying life instead of constantly dwelling in the past or worrying about the future.

A love of reading and writing is the hallmark of erudition. Whenever there is a toss-up between whom to hire, work with, fund, make friends with, promote, or do business with, the person who is a more skilled reader and writer will usually win. The reader and the writer are the most likely to succeed in entrepreneurial pursuits, sales, and any business that requires working face to face, or where comprehension and writing are even remotely crucial. The corollary of these practices is a natural improvement in speaking ability, which nicely compliments the array of competency triggers.

Now let's jump into the strategy and action section where we'll build some realistic habits, practices, and experiments around reading and writing.

Strategy and Action

"On this path, it is only the first step that counts." –St. Jean Baptiste Marie Vianney

"…less than 10 percent of the people who buy a book read past the first chapter. At first, I couldn't believe that statistic." – Anthony Robbins

Learning That Compounds

We can start to build the reading habit by only reading what we love. Most of us have been taught to hate reading through boring assignments in K-12 and college. So after we firmly reestablish and foster an enjoyment of reading (around 66 days of consecutive practice), then we can work to strategically expand what we read. The challenge or baseline I set for myself is to always read more than 52 books in a year.

At first this was a bit of a stretch for me, as I didn't have a reading habit developed. I made sure I didn't quit before I had given it a full chance. So, when I first started the 52+ books a year challenge, I started embarrassingly small. I made it absurdly easy for myself to win. Some of the first books I read and counted towards my 52 were anywhere from 14-60 pages in length. They were short stories, essays, novellas, and I counted them all to build up the practice.

As I chalked up these small wins, my enjoyment of reading increased. I started to throw in some longer books and started to add books in subject areas I was curious to learn more about. It was only a short time into my first year of going for the 52+ that I started seeing the benefits. My general anxiety lessened, my mind continually felt at ease, I was more creative than ever, and I avoided the trap of going to graduate school.

The effects of the 52 books a year challenge created a positive feedback loop. The more I read, the more ideas I got, and the more value I delivered in my interactions with others. Once I felt the benefits of this habit, it became easier to continue. Those benefits compounded and the cycle continued. I completed reading the 52 books a year challenge with around 65 books finished. The next year I set my goal higher and have been enjoying the benefits from it ever since.

The most important idea when starting this practice is to make it easy to win. Feel free to tailor or adjust this challenge however you see fit. You only need to do this practice for a single year and, viola; you've read more than 95% of Americans ever will.

Don't burn yourself out too quickly in the beginning! If 52+ books a year feels daunting, try 30, 24, or 12. The idea is to pick something which feels a bit beyond your reach, and then slightly increasing the goal after you reach it.

Once you set your goal, it's important to either make it known or set up some kind of accountability. Does anyone know you're undertaking this challenge? Is there anyone you can begin this challenge and trade books with? When we get creative about constructing the right incentives and stakes, it's hard to not to achieve what we aim for.

To help get you started with this challenge, at the end of this section there is a list of 65 highly curated books.

Writing Strategy and Action

"The purpose of the first line is to get people to read the second line."
–David Ogilvy

The global proliferation of Smartphones and tablets is accelerating. As millions of these screens have more eyeballs on this limited real estate, skill at writing is increasingly in demand. Anyone can write, but few people can get their point across clearly.

Most people never begin to improve their writing. They get hung up on punctuation, grammar, prose, and voice. These all have some varying degree of importance, but they're not really that important. All those aspects of writing can be learned and improved rapidly. But they can't be improved until we give ourselves permission to be bad at them publicly. We only become better at writing by exposing it publicly to a marketplace of people who don't know us. The hardest part of getting started at improving our writing is overcoming self-censorship. There are layers of culture, peers, and society which we have to scrub off in order to translate meaning and emotion into our writing.

The intentions behind what we write are important. When in doubt, our intentions to write and communicate well matter more than anything else. By striving to convey our intentions, we'll often find ourselves improving. The next most important training item to build writing is interesting experiences. Are we executing on our fridge lists? Are we continuing to find interesting people to meet and talk with? The person who has had an interesting life full of risks, ups, downs, close calls, ambition, false starts, and success, is the person who will have writing worth reading. Grammar, editing, punctuation, and prose are all secondary to having something valuable to convey.

Getting started at trying to improve my own writing came only after I started reading more. I started by trying to improve my writing in emails, and then expanded the practice into getting to just a few hundred words a day. From there I expanded again. Improving our emails is a great place to begin, and below is a practice which we can to apply to any writing of any length.

There is an old trick copywriters use, called AIDA. It stands for outlining a short message (such as an email) using: Attention, Interest,

Desire, and Action. Next time we're writing emails, take a post-it note, write out AIDA, and seek to write your email into the confines of that acronym.

Many people think that AIDA originated to sell things, and they're right. But thinking this is negative, or selling is sleazy, is pretty foolish. Selling implies conveying, and then delivering, value. In order to sell things over the long term, it's mandatory to consistently provide value and deliver on our promises. Remembering AIDA is a way of delivering value in a way which respects other people's time. Think about the last time somebody sent an email which made you smile, or was an enjoyable read. That's a person who cares about you and values your time. The person who sends us a rambling two- page email they've forced into our inbox doesn't value their own time, so there is no way they can begin to properly value ours. Nothing says, "I value your time and want to provide you value," than writing email messages in the AIDA format. So let's say you're writing an email to a significant other. Try writing an email to really capture their attention.

The AIDA format example:

Scenario: I'm writing an email to my wife.

The Email Text

- **A**ttention- Subject: Throwback Thursday
- **I**nterest- Remember when we went to Waterfront Pub on Q Street last summer? It was on a Thursday.
- **D**esire- It's been awhile since we've been on a date; we should go back there…
- **A**ction- Let me know if you can be home at 6:00 pm tonight and we'll go.

-Me

Scenario: Say you've done some research and there is a job you're interested in landing. You are reaching out to the owner of the company, the hiring manager, or the hiring manager's boss.

Always remember, never reach out to the HR recruiter; at most places there is no incentive for the HR recruiter to help you land a position. They are only motivated to do exactly what their boss tells them, unless it's an incredibly entrepreneurial organization.

Below is a sample email you could send to the CEO or hiring manager in reference to a specific position you're interested in, or if you're trying to get your foot in the door.

Attention Subject: (a solution for something they need)

Interest- (Hi Name), (Your name) here with five ideas to expand your distribution for (their product or service here). I think these channels are currently being overlooked because (a reason showing you've done research here).

Desire- I believe these five ideas would be perfect for expanding your distribution and increase your revenue because (a reason which demonstrates you understand their business and you've done research). (Write five ideas here.)

Action- I would enjoy implementing these ideas to increase distribution and revenue for (their product or service). If you're free at 3:30pm EST on Tuesday or Wednesday, I'd love to chat briefly to learn how I can execute the ideas listed above!

Best,

-Your Name

This idea is specific, but the general theme is to practice reaching out with the AIDA format and deliver value. Value can come in the form of ideas demonstrating your competence, connections, customer referrals, or anything which is helpful and shows you are good at learning and creating value. This practice is also a phenomenal way to demonstrate another competency trigger – giving value before asking for value.

The idea of practicing our writing through AIDA and better emails is just a small, simple step. Something as small as writing all your emails in the AIDA format consistently every day, can set us up to be leaps and bounds ahead of any co-workers or professional competition we might face. If something really inspires or frustrates us in the coming months, feel free to break out and write an opinion editorial or analysis of whatever challenge is going on. Emotions are good starting points for practicing writing.

The Best 65 Books For Aspiring Entrepreneurs

"Civilization is in a race between education and catastrophe. Let us learn the truth and spread it as far and wide as our circumstances allow. For the truth is the greatest weapon we have." –H.G. Wells

"A reader lives a thousand lives before he dies. The man who never reads lives only one."
–George R.R. Martin

65 Books to Spark the 52+ A Year Practice

This list of 65 books is a starting point. Whether you get started with reading 52, 24, or 12 books a year, feel free to pick, choose, and discard based on your interests. Or, if you're already an aggressive reader, feel free to pick and choose any of these titles to supplement your existing queue.

Incerto is the name for a collection of books by Nassim Taleb which make up the first four books on this list.

1. *Fooled by Randomness*
2. *Black Swan*
3. *Bed of Procrustes*
4. *Antifragile*
5. *Zero to One* by Peter Thiel
6. *The Hard Thing About Hard Things* by Ben Horowitz
7. *Unbeatable Mind* by Mark Divine
8. *The 48 Laws of Power* by Robert Greene

9. *The 50th Law* by 50 Cent and Robert Greene
10. *The 33 Strategies of War* by Robert Greene
11. *Mastery* by Robert Greene
12. *Mindset: The New Psychology of Success* by Carol Dweck
13. *Average is Over* by Tyler Cowen
14. *So Good They Can't Ignore You* by Cal Newport
15. *The Tyranny of Experts* by William Easterly
16. *As a Man Thinketh* by James Allen
17. *The Four Hour Work Week* by Timothy Ferriss
18. *The Four Hour Body* by Timothy Ferriss
19. *The Four Hour Chef* by Timothy Ferriss
20. *Love Yourself Like Your Life Depends On It* by Kamal Ravikant
21. *Live Your Truth* by Kamal Ravikant
22. *Think and Grow Rich* by Napoleon Hill
23. *Choose Yourself* by James Altucher
24. *Rise of Superman: Decoding the Science of Ultimate Human Performance* by Steven Koteler
25. *Abundance: The Future Is Better Than You Think* by Peter Diamandis and Steven Kotler
26. *The Rational Optimist: How Prosperity Evolves* by Matt Ridley
27. *Economic Facts and Fallacies: Second Edition* by Thomas Sowell
28. *Drive: The Surprising Truth About What Motivates Us* by Daniel Pink
29. *The Talent Code* by Daniel Coyle
30. *Steal Like An Artist: 10 Things Nobody Told You About Being Creative* by Austin Kleon
31. *Jonathan Livingston Seagull: The New Complete Edition* by Richard Bach
32. *Illusions: The Adventures of a Reluctant Messiah* by Richard Bach
33. *The Obstacle is the Way* by Ryan Holiday
34. *Wool* by Hugh Howey
35. *How to Win at the Sport of Business* by Mark Cuban
36. *The Prince* by Niccolo Machiavelli
37. *Dune* by Frank Herbert

38. *Atlas Shrugged* by Ayn Rand
39. *Anthem* by Ayn Rand
40. *The Virtue of Selfishness* by Ayn Rand
41. *The Sticking Point Solution* by Jay Abraham
42. *The Start-up of You* by Reid Hoffman and Ben Casnocha
43. *The Slight Edge* by Jeff Olsen
44. *Intellectuals and Society* by Thomas Sowell
45. *Psycho-Cybernetics* by Maxwell Maltz
46. *The True Believer* by Eric Hoffer
47. *Basic Economics, 4th Edition* by Thomas Sowell
48. *Meditations* by Marcus Aurelius (Gregory Hays Translation)
49. *Turning Pro* by Steven Pressfield
50. *The War of Art: Winning the Inner Creative Battle* by Steven Pressfield
51. *Man's Search for Meaning* by Viktor E. Frankl
52. *Letters from a Stoic [Annotated]* by Lucius Annaeus Seneca
53. *The Flinch* by Julien Smith
54. *Spark* by John J. Ratey
55. *The Education of Millionaires* by Michael Ellsberg
56. *Uncertainty: Turning Fear and Doubt into Fuel for Brilliance* by Jonathan Fields
57. *Self Reliance* by Ralph Waldo Emerson
58. *Emotional Intelligence* by Daniel Goleman
59. *Free to Choose: A Personal Statement* by Milton and Rose Friedman
60. *Little Bets* by Peter Sims
61. *Secrets of Power Negotiating* by Roger Dawson
62. *Super Intelligence* by Nick Bostrom
63. *Shantaram* by Gregory David Roberts
64. *Good to Great* by Jim Collins
65. *The Black Book of Communism: Crimes, Terror, Repression* by Jean-Louis Panné, Andrzej Paczkowski, Karel Bartosek and Jean-Louis Margolin

14

Mastery of Technology and Once You Go Amish...

"There are many things of which a wise man might wish to be ignorant" –Ralph Waldo Emerson

"Humans are the reproductive organs of technology. We multiply manufactured artifacts and spread ideas and memes." –Kevin Kelly

Rules, Threats, and Upside of Technology

We thrive when we use technology as our slave, and suffer when it becomes our master. This chapter is about becoming a master and steward of technology. You probably know what they say:

Once you go Amish on adopting technology, you never go back! Maybe they don't say that… But anyway, consider this story:

Somewhere right now, in upstate New York, there is an Amish community. This community is thriving. They're enjoying a boom from the city folks who have become obsessed with organic vegetables, produce, and handmade furniture. Every weekend these city types arrive and flood the Amish community with money.

Meanwhile, the Amish community lives simply and prudently. They save all the money they earn; they never go into debt. Recently, they have begun combining their income to aggressively purchase more and more land in upstate New York. As the cities and suburbs expand outwards, they quietly purchase huge swathes of valuable land. One of the Amish's rules about land they purchase is—they never sell it. Land is becoming scarcer, and the Amish are strategically capitalizing on this fact.

Land acquisition isn't the only thing the Amish do that's fascinating. They also have rules and systems set up for how they use and adopt technology. Most people think the Amish are luddites, or don't use technology. This couldn't be further from the truth. The Amish have rules set up to carefully test and adopt the right types of technology. They are careful to make sure they understand the tradeoffs of new technology before bringing it into their communities. The Amish understand that comfort can bring diseases and iatrogenics.

One technology the Amish are currently testing out in their communities is cell phones. To make sure they don't poison the community, they do small incremental beta tests of technology, then reflect on the pros and cons before large-scale adoption.

I can see the town hall meeting now:

"Jebediah! Why do you insist on needing a phone?!" asked Hamish, the town elder.

"Brother Hamish, I need a phone so I can better communicate with Mordecai. Our handmade furniture business is booming. We need to speak daily and he lives a mile away!" said Jebediah.

"Jebediah, do you not see all the children from cities who come to our roadside shops? You see them worshipping these phones! Their faces are pasty white; they tap incessantly on those devices!" roared Hamish.

"Brother Hamish, you are correct. These pasty whities do worship their devices with devotion! But the parents of these device worshippers bring us money which helps us acquire more land!"

"Indeed the whities pockets are loaded," nodded Hamish approvingly. He motioned with his hand for Jebediah to continue...

"I have an answer to help us reap the benefits of these phones, without all the hunched tapping," said Jebediah slyly.

"Jebediah, I hope what you propose is good. I fear you want to put Smartphones and Candy Crush in the hands of our children!" said Hamish.

"Brother Hamish, I've found a way to prevent the device worship. The answer is in the perfect mobile phone technology these city folks have already discarded. We can acquire these cheaply. The answer is *Clamshells*!"

"Jebediah, YOU KNOW I'm allergic to seafood!" boomed Hamish.

"No, Hamish, these *Clamshells* are some of the first cell phones. We can test these safely in the community. Let the community watch how these *Clamshells* aid Mordecai and me. If they carry evils, then let the community judge!" said Jebediah.

"Very well then! Who is in favor of allowing Jebediah and

Mordecai to test these *Clamshells*?" asked Hamish.

"AYE!" roared the crowd.

So obviously, this is part joke and part truth. There is some serious value to observing how the Amish adopt technology. The Amish are actually in the middle of considering, and I believe testing, cell phones. The rules the Amish have set up for adopting technology might be limiting, but at least they have some rules and guidelines set up.

Portals to Anywhere, Everywhere

"The internet makes dumb people dumber and smart people smarter." –Kevin Drum

Everywhere around us in modern American culture are bright and shiny machines. All different types of technology are becoming cheaper and more ubiquitous every single day.

We've seen that technology is anything which allows us to do more with less. By itself, technology is neutral; it doesn't do much of anything until we interact with it. It's up to us to make sure we interact with it in a way which increases our abilities and productivity, and not the reverse.

Most people have no heuristics, rules, or tests set up to assist in deciding what technology to adopt and at what cost. Adopting technology prudently doesn't have to be wildly difficult. It's as simple as being honest and objective when we look at how we spend our time interacting with it.

Once we have an idea about the pros and cons of what technology provides (and they are not all pros), we can selectively choose the specific types we want to adopt. The Amish are notorious for being what appears to us as "slow" adopters. What appears as slow is really just methodical. We should feel free to do the same, perhaps turning off notifications on our phones completely, or refusing to keep no more than a handful of applications on our phone at any given time.

Most technology is amazing, with few to no tradeoffs, but by relying on some technology too much, we can become soft, idiotic, and little more than drones tapping screens for quick hits of dopamine.

The lenses through which the wonders of modern technology are best viewed come from two places: the Uber (no pun intended) successful technology entrepreneurs who are rebuilding our country, and the Amish. By blending these two approaches, we're light years ahead of the majority of our country and the world.

The goal of this section is to remember one core heuristic: we should be masters of technology. "Master" in this case means being able to select, utilize, control, and adopt technology with some sort of reason as to why we are doing certain things. This is perhaps a lofty goal as we look around and see people hunched over their phones. Even when we're driving in the car, there is a good chance if we look around, we will see the person in the next lane is only partially interested in the road. The people driving with their phones over the wheel are unknowingly seeking quick hits of dopamine from the bright lights and notifications of their phone. This is a complete demonstration of addiction and stupidity; they're seeking to get their fix at the risk of killing someone with the vehicle they're supposed to be driving.

All around us are examples of people being used by technology. Ever since the first televisions came into American homes, people have talked about how it would bring in wonderful educational content. Millions of hours later, that's definitely questionable. Many families went from never having seen a television to watching 4-8 hours every day. Technology has started to control, influence, and use many more people than we might realize.

When we define technology as "doing more with less" it

purposely leaves the definition of "more" up to the individual. It's up to us to decide what we want "more" of. This requires work, strategy, and setting up systems on our part. This definition of technology is important because it isn't limited to electronics. Technology is any innovation which helps humans do more. Some examples:

- A chair is technology.
- A Smartphone is technology.
- Our bodies could be thought of as a technology.
- Applications for technology might be a handle on a knife or wheels on a suitcase.

In the case of evolution and our human bodies, consider the killer application of opposable thumbs. This is an application on top of a technology which helped us win early evolutionary battles against other species. We have technology and we have applications which run on technology. It's up to us to never find ourselves slaves to technology, only masters.

Skills for Mastering 21st Century Technology

Those who master 21st century technology will use language to better define and set parameters. The following ideas and concepts can better direct our use of technology: Curation, Filtering, and Synthesis.

Curation: Personally regulating, adding, dismissing, or testing the technology we use so we exercise mastery and awareness over what it brings to our daily lives. This might mean we have a Kindle to read with no apps on it to ensure it's used only for what we intend.

The most important idea behind curation is that in order to justify the use of a new technology, we should be able to articulate what that technology is allowing us to do and how it is helping us achieve what we want.

The second idea behind curation is being aware of the tradeoffs and downsides of using new technologies. The most glaring example of why we need to curate technology and how we use it are the many young people in America developing literal hunchbacks from bowing to view the screens of their phones. This is a prime example of being used by technology.

Filtering: After we thoroughly consider which technologies we will use and why, we then selectively use existing technologies to create barriers to prevent us from being used or taken advantage of by the downsides of technology.

By introducing the mindset of filtering, we can develop heuristics such as "never pick up a portal to everything without specific intent" to better control our impulses to mindlessly pick up our Smartphones when we're bored.

We can also use simple filtering mechanisms on our computers such as Rescue Time to turn off the Internet while we are working. We can use filters such as Feedly to consciously select content and put it in a single feed. These filters can prevents us from wading out into the noise of the Internet and getting distracted.

People as Curators and Filters

One of the benefits of living in an interconnected world is that the best content is being curated for us. There are people, websites, and influencers who are motivated to help find the best and most useful information and present it to you. This is an immense benefit. By finding trusted sources for content relevant to us, we don't have to continually expend extra resources sifting through old mainstream news and information sources.

Why does developing a mindset primed to filter what comes

through our digital devices matter? It matters because, in any given day, we have a fixed amount of willpower. Dr. Kelly McGonal, in her book *Willpower*, explains that as we encounter distractions throughout the day, we are slowly decreasing a finite amount of willpower. This means as we distract ourselves and the day wears on, we slowly degrade our ability to concentrate and do what matters. By using filters and curation sources, we can safeguard our willpower for what matters most.

Synthesis: We want to be actively thinking about what we're creating with the technology we're using. Are we taking full advantage of it? Are we using it to fuel new creations? Or are we using it to consume and distract ourselves?

Synthesis is the process of using technology to help zero in on the resources and information we need. It urges us to find specific tidbits of knowledge we can use to plug gaps in our skills which will help us reach our goals. The goal is to synthesize every interaction and piece of knowledge we have with technology to direct them towards the outcomes we want to create.

An example of synthesis might be if we want to find a job or career in a business which is rebuilding America. We might pick a company which is helping to rebuild America from the list in Chapter 15.

Then we would use the list of connectors in Chapter 16 to determine the best way to get there from our current level of skills.

We would use the strategy and action section of this chapter to set up the necessary social profiles to establish credibility, then start reaching out to decision makers at the companies we're interested in working for. We can ask them what specific skills are needed to land a position there. Then we can use technology to learn that skill or credential and apply.

Along the way in this pursuit, we might use an app like Feedly. This app allows us to select the blogs, industry news, and publications we read, and bring them all into a concise layout to browse at our convenience. Feedly is free of advertising and we don't need to venture out onto the web or encounter endless advertisements. Next, we might use Twitter on our phone to comment or add value to the discussions and content around the industry we're interested in. This is an example of curation, filtering, and synthesis in action to ensure we use technology, instead of letting it use us.

Strategy and Action

"It's hard to say exactly what it is about face-to-face contact that makes deals happen, but whatever it is, it hasn't yet been duplicated by technology." –Paul Graham

"Blow up your TV, throw away your paper..." –John Prine

Harnessing What We Create

Assuming you're not from an Amish community, you will probably have access to computers and a Smartphone. But if you ARE from an Amish community, you'd better be reading this book via paperback and not on a Kindle!

Let's assume you're not, and you actively use technology in your daily life. This strategy and action section offers some ideas to help better curate, filter, and synthesize what we choose to use. Feel free to use these ideas, discard them, or put your own spin on them.

Television Mastery

The idea to master our use of television starts with placing a buffer between us and the content from our television. Cable TV is largely a gateway to reruns, commercials, "news," and crap. By eliminating cable, we effectively set up a buffer between our minds and trash. Now we can take control of the technology and add an application to our television to synthesize what we take in. An example might be to connect our television to the Internet or link it up with Amazon Fire TV. Now we have a barrier where we're forced to make a conscious decision to select what to watch instead of just turning on the television and consuming.

Computer Mastery

Rescue Time: This app lets us turn on and off the Internet on our computers for a pre-determined length of time. I like to use Rescue Time to turn off the Internet on mornings like today, when I'm writing and don't want the temptation to start looking things up online to interfere with my work. Rescue time helps filter out the noise which would degrade our willpower first thing in the day. When we control the Internet on our devices, we make conscious choices about what and how we'll use it.

Computer and Smartphone Mastery, The Game Changer

The most valuable way I've found to control and drive my use of a Smartphone and computer is as follows:

- At night before bed, or in the morning, I create a to-do list for the day on a 4x6 note card.
- From that list, I will select or write in the three most important things to do. The idea is that if I got nothing else done, and only completed these three things, I would still make progress towards an outcome I want to create.
- I write them down, circle them, or annotate them.
- The next day before I do anything online, before I even TOUCH my Smartphone…
- I do those three things. No outside stimulus until they're done.
- This is a simple way to get where we want to go, without derailing ourselves or chipping away at our willpower first thing in the morning.

This is an act of defiance towards always being connected, it's a way to refuse to be controlled and ensure we use technology efficiently.

Content Curation and Filtering Online

Feedly: This app brings all the feeds from websites into one place where you can read or share select articles and content. Simply login and add the specific sites you want to get news and updates from. The app doesn't bring in advertisements so you're able to read things without constantly seeing display advertising, which degrades your willpower. Once you have Feedly set up, you can add the app to your browser toolbar, and add additional sites accordingly. If you're learning something, or trying to break into a new industry, simply add some sites of curated content from influencers in that space. Over time, you can use Feedly to build a custom stream of highly relevant, amazingly useful content which will save you hours of time.

Upside Exposure: The best and most scalable way to increase our exposure to opportunities is to set up online social media accounts on Twitter and LinkedIn. Depending on your industry, using Medium might be helpful but we don't need to worry about that for now. These accounts allow us to capitalize and create opportunities. If you don't have a personal website at this point, don't worry. LinkedIn can hold your resume and be the hub of your online presence, and Twitter is just a more specific lead generator and way to interact with people.

An opportunity generator for ourselves is something we set up which can advertise and promote us while we're not directly working. In business, we might call the creation of these opportunities "lead generation." In our personal lives, many people just call this smart. Many people have grandparents who advised them to never look for a job, but always look for an opportunity. Having social profiles set up to work on our behalf to uncover and expose us to opportunities would make our grandparents proud.

Platforms as Opportunity Generators

By taking a small amount of time to perfect these opportunity generators, we'll be able to dive into the strategies and action sections in chapters 16-19 and start increasing our personal exposure to uncapped upside. The idea is to interact with others, send cold emails, write articles in our industry, and link them back to the platforms containing our resume.

Twitter: You don't even have to tweet, but by having this set up you'll be able to claim your name and prepare for when you do tweet. Just sign up and post a single tweet. This helps in case anyone on Twitter searches for your name, and this single tweet can direct interested parties to the appropriate place:

"Thanks for stopping by! Too busy to tweet at the moment, but let's connect (link to your email address or your LinkedIn profile)."

The link to connect can go to the places we mentioned above, or to your profile at your current company, your personal website, or even a short 30- to 60-second video introducing yourself, your skills, and if you're a job seeker, what you're looking for.

LinkedIn Profile: If you're looking for a job, opportunity, or promotion, this is *the* place to start. Recruiters scour LinkedIn all day long looking for applicants. There are a million free resources online showing how to optimize your LinkedIn profile, so I won't bore you with details here.

The idea is to set up a LinkedIn profile and, over time, add more than 500 connections. The LinkedIn publishing platform is still in its infancy, and it's a great time to use it to begin writing industry-specific articles. The publishing platform will likely grow even larger and more esteemed as time goes on. Getting involved now and having a public place such as your LinkedIn profile to write and become noted in your industry is a great way to generate opportunities and serendipitous connections.

If you need a job, make sure your LinkedIn profile is cleaned up and your resume is posted. Then, on LinkedIn's publishing platform write a post titled something along the lines of "Ten Things I Didn't List on My Resume". Include everything, including stories or decisions which have built your character.

You can follow up that post with "10 Reasons I'm Ready to Work At (company here)", citing what you bring to the table. Or something like "10 Ways For (company here) to Increase Their Sales". The idea is to stand out from the traditional channels of resume spamming and find opportunities before they're advertised to the market.

LinkedIn's expressed long-term goal is to turn up contextual economic opportunities for the right people, at the right time. If you don't understand what this means, look it up and write a post about why the long-term vision of LinkedIn is important. If you get it right, LinkedIn will likely want to promote your writing. If you're a horrible writer, record what you want to say and have it transcribed, or hire an editor on oDesk to proofread your work. LinkedIn is *the* platform which can unlock thousands of opportunities.

Personal Website: This isn't needed at first, but for specific people this could make sense. The credentials of the future involve being able to cite a portfolio of work at a moment's notice to show a potential employer, ally, or friend. A personal website, using an incredibly cheap and full-service solution provider like Squarespace, is perfectly suited to showcase our work, projects, and the skills we have.

Launching a personal website is also a great excuse to grab the domain name for your name. Most .com domains will be taken, but there is a good chance you might be able to find a .co or suitable alternative such as (www.YourFirstName-YourLastName.com). If you want to blog or start a business and sell products or services from a personal site, Squarespace lets you sell digital products and services using the awe-inspiring payment processor Stripe. Stripe will soon be accepting Bitcoins.

A Few Notes on Platforms

Ensure congruency across platforms. That means make sure all your social media profiles display the message you want to convey.

For the advanced folks, or those who already have LinkedIn and Twitter accounts set up and are using them well, you can consider other opportunity generators such as Medium or AngelList.

AngelList will be incredibly useful as we dive into reverse engineering entry into high-growth businesses in the next chapter, so consider setting up a profile there.

Medium: This is where you can write about certain industry- or life-specific topics and give those writings exposure to a large audience. This is an alternative or supplement to publishing on LinkedIn, and might be a bit more useful to those interested in startups, technology, design, art, or writing.

AngelList: This is where you can learn all about startups and technology businesses, and discover how to procure jobs with them. If you're a qualified Angel Investor, you can start looking for companies, funds, or syndicates for investment. Anyone interested in technology should become familiar with this site. AngelList is the best platform on the Internet to connect early stage businesses with accelerators, angel investors, and incubators. They're also building index funds for private companies. Businesses which are rebuilding America will be listed on AngelList.

All these platforms we have mentioned can become opportunity generators if used appropriately.

Once you're set up, at least on LinkedIn and Twitter, you've embraced the mindset of exposure. Now when you start seeking out opportunities, you have an online presence to capture opportunities and introductions as you go.

15

Reverse Engineering Entry Into Fields That are Rebuilding America

"During the past three decades, startups in the United States have created nearly 40 million American jobs, all the net job creation in the country over that period." –Steve Case

Ethics or Sugar Water?

Consider what America has brought the world: technology, cures for diseases, and more wealth in a few hundred years than in the entire history of humanity. We were the first nation to reach the moon. We defeated communism, or at least survived long enough for communism to consume itself. We will be the platform from which human life becomes multi-planetary. We are living in a rare and unique moment in history. We can ally ourselves with companies doing amazing work, and we don't have to put aside our ethics. We can join with those that are changing the world and aren't relying on government bailouts to stay afloat. There are businesses built by risk takers which need to hire great people. We are the class of people they need.

There is a story about Steve Jobs approaching John Sculley asking him to come work for Apple. At the time, Sculley was an executive who worked at a large soft drink company. He first laughed at the job offer from Jobs. Sculley was incredulous that Jobs would even ask him to come to work at Apple, then just a four-year-old company which had been started in a *garage*.

Jobs listened to Sculley's answer politely, and then stared at him before responding. Jobs asked Sculley, "Do you want to sell sugar water for the rest of your life or do you want to come with me and change the world?"

Working for a business or in an industry changing the world is the most ethical thing we can do with our energy. The greatest virtue doesn't lie in philanthropy. The greatest virtue lies in discovering where we can create wealth without hurting anyone else; where we can create opportunities for all of humanity.

Who and Where are the Builders?

If there was a group of companies taking a small amount of money and turning it into a large amount of money, we'd know about them, right? I'm not talking about the companies that make the average news headlines. Many of those companies are okay, but I'm talking about the thousands of companies collectively driving American progress. In our country, we have a group of small businesses, startups, and venture-backed technology companies which account for nearly all the technology, jobs, and progress we create.

We would think there would be a race to work at these innovation hotbeds. But in truth, working at these startups and businesses requires enduring some risk. This risk is nothing compared to what the founders of these startups have endured. But there are many people in our society who would rather find jobs which require others to personally bear risk in order to keep them risk free. These people are cowards. Veterans of an all volunteer military are the exact opposite.

As veterans, we've stepped up once and have borne significant physical risk for our country. I can't help but wonder what would happen to the American economy if the 2.8 million post 9-11 veterans all aligned themselves with startups creating value. What would happen if we all voted for American progress by joining this ecosystem of entrepreneurship? Whether it's by working at a startup, or even a larger technology company, then launching our own ventures when the time was right, this would create an economic boom in America. This flight of veterans into technology and startups (the only net job creators of the last three decades) would do more for the American economy than any politician ever could.

Consider the staggering facts about venture-backed startups in the United States. The amount of money invested into them every year is a tiny .02% of GDP. This means venture capital funds invest around $346.6 billion into a portfolio of startup investments every year. With that investment, they create around 11% of all private sector jobs. That's around 15,378,000 jobs! Venture-backed companies have created revenue which accounts for over 21% of American GDP, or around $3,633,000,000,000 (3.6 trillion dollars!). The top 12 largest technology companies in the U.S. account for ~$2 trillion of our GDP. Those dozen largest companies were all venture backed. Venture-backed companies each start with a few hundred thousand or million dollars and in turn spread innovation, jobs, wealth, and opportunity.

Now for a stark comparison, consider the job creation impact of the 2008 Federal Stimulus plan. This plan directly cost taxpayers around $800 billion of government debt and borrowing of money we do not have. This means limiting the options of current U.S. citizens and mortgaging the future of unborn children. Depending on reports ranging from think tanks to the Congressional Budget Office, it's estimated that the Stimulus "created or saved" between 0 and 1.5 million jobs.[29] Those numbers are disputed, but without more government debt, those "jobs" which were "created" don't continue.

Venture-backed startups aren't investing with debt. They're investing with money given to them voluntarily. Across the entire country, these startups are taking a fraction of capital and producing returns thousands of times greater than government deficit spending.

For the first time in history, we have opportunities to work at the type of companies which are taking a little bit of capital and turning it into massive amounts of value. In fact, it's these startups and industries which have the potential to rebuild America and ensure humanity becomes a multi-planetary species.

As Americans, we're blessed to be able to choose work and careers from the top of Maslow's hierarchy of needs. We can choose the type of work which causes us to wake up excited and go to sleep exhausted and satisfied from a day well invested. We can choose the type of work which gives us exposure to risk, but where we create innovations which make life better for every person on the planet. We don't have to choose small, when with a little effort, we can do our homework and choose to work on pursuing incredibly audacious and long term ideas.

This is the path where we can speed up our entrepreneurial learning. This is the path where, along the way, we'll meet life-long friends, and co-founders, mentors, and investors. We'll learn on the front lines of capitalism. Remember, capitalism's true nature isn't about competition, but creating value in a unique and differentiated way where for a time, the business holds a monopoly. It's these brief (sometimes decade-long) monopolies which must be continually won for our country to prosper. This is a cycle of not only creative destruction, but one necessary to produce innovation and avoid stagnation.

The path of joining this experiment and finding work we love to do is simple. But most people will not choose to do it. They'll start by looking at whatever jobs are listed; they'll look for what's secure, for what has a great salary. They won't think long-term or about what their equity options are. They'll pick randomly instead of acting tactically from the beginning.

In order for America to continue to lead the world, we as individuals must be willing to take risks in our professional lives. The best opportunities (what our grandparents called jobs) are rarely advertised. They are secrets waiting for us to discover them. So where do we find these opportunities?

There are three core places where we can identify and find the right startups to medium-size businesses, and venture-backed technology companies. This is where the best near and long-term opportunities lie for veterans.

Three Places to Look for Opportunity

1. The *INC 500* list can be found every year in a specific issue of *INC* magazine. This list is especially useful if you're looking for something outside of traditional technology or software companies. This list contains rapidly growing companies located all over the United States.

2. AngelList, Angel.co: Build a profile and start learning. This is the future of angel investing and job opportunities at startups, and is the way to connect with the right investors through competency triggers, signals, or referrals. A competency trigger on a platform such as AngelList would be a completed profile which looks great and has many impressive connections, or a stellar work history. Signals that you're ready for investment might be a company page showing proof of concept, traction, or who is investing in what round. Referrals would be introductions to the right people at the right time. This is just a small sample of all the benefits AngelList provides, and it's definitely one of the best places to start doing your reverse engineering homework and opportunity creation.

3. The Portfolio tabs/pages of top-tier Venture Capital firms. These pages contain lists of all the companies a particular VC firm has invested in. Some of the top tier VC firms in the United States include:

- A16Z
- Greylock
- LowerCase
- Founders Fund
- Foundry Group
- Google Ventures
- First Round Capital
- Upfront Ventures
- True Ventures

Now let's get tactical about aligning ourselves with the growth of these economic engines.

In the Mastery of Technology section, we reviewed the platforms on which we can create accounts in order to reach out, learn, and create opportunities for ourselves. Once we have these social profiles and platforms launched, we can begin reaching out, meeting people, learning, delivering value where we can, and getting answers from the front lines. This is where reverse engineering entry into these fields becomes more than a possibility. By being active on these platforms and in emails, we get a chance to make it a reality.

After we reach out, in the next chapter we'll see how we can create the credentials we need to enter and win opportunities at the companies we're interested in. Maybe the credentials we need are a body of work: a portfolio of apps, demonstrated coding ability via a GitHub portfolio, a design portfolio via Behance, sales numbers from our last job, or testimonials regarding our work from people the decision maker or founder at the startup knows and trusts. The point is, we can figure out and earn the credentials we need later. The more important task is to get the lay of the land about who is rebuilding America.

Reverse Engineering Entry Into the Businesses Rebuilding America

- Learn about many of the companies through the three main hubs for them: INC 500, AngelList, and VC Portfolio Pages.
- Find 30-50 startups/companies which fascinate us. Study everything about them including what they sell, the markets they're in, the stage of business they're in, what jobs they need to fill, what products they make, and who makes up the founding team. The more knowledge we have about them, the better.
- Start a correspondence with some of the founders of these companies. Reach out to the people who hire and make decisions. Reach out to the sales, engineering, growth, or design team, depending on where your specific interests lie. Introduce yourself and determine how you can be helpful.
- Determine the exact type of skills we would need for them to hire us immediately. Do we have them already? Do they offer a trial period of working for them before being considered for a full-time hire? What can we do or what can we show them to make them unable to ignore us? It's our mission to find out. Are they looking for a college degree? Are they looking for candidates from a developer boot camp? Find out the exact credentials and deliverables you need to get hired. If coding or designing is out of the question for us, the best startups always need sales people and amazing customer services representatives. The first person answering the phone at Facebook isn't answering phones anymore.
- From the intelligence we gather, we can then acquire that training, or start building a portfolio of work and credentials based on the

unifying theme of what all the different people we asked said was needed. So ask 30-50 people what exactly we need to get hired, find the unifying themes and specific credentials they continually cite, and then find the best way to get that training. As we go about getting this training, we can go back to them from time to time to keep them apprised of our progress, and continue to hustle for an opportunity to get our foot in the door. In the next chapter discussing College or Not, we'll see how we can ensure any training we pay for offers a great outcome.

- Continue the process of meeting individuals, founders, and key employees at these businesses. Connect with every one of them on LinkedIn after the initial conversation. Find the real scoop on what's going on in their industry. Learn even more about these industries. Write articles showing you understand specific nuances and services these companies offer. Promote these companies via articles or interviews. Send these articles and interviews to the founders you want to work for.
- At this point of reaching out to all these different types of founders and startups, we'll either find ourselves with a job or with the intelligence from the front lines which will point us to the credentials (skills) we need to acquire.
- If we've landed a job by this point, great! If not, and we need to get more training, simply keep in touch and keep these people updated on your progress. Hardly anyone in business (or life) is tenacious enough to follow up. If we follow up with the contacts we've made, we'll be ready at the moment our skills match what they need. This is when we discover or are offered jobs before they're advertised.
- Now for the training part of the equation. Do we have to use college to get there or is there a quicker path?

Strategy and Action

"If someone offers you an amazing opportunity and you're not sure you can do it, say yes - then learn how to do it later." – Richard Branson

"Fear is the disease. Hustle is the antidote." –Travis Kalanick

Crushing Myths

There are a few myths which prevent many people from getting started, or even trying to reverse engineer something amazing. If you don't need to review these myths, feel free to jump to the next chapter. The myths below are many I've heard repeatedly, and they can creep into just about anyone's thinking:

Myth: I need an Ivy League degree (or some other amazing credential) to unlock opportunities.
Truth: It doesn't matter what you think you need, or if some people say college is worthless while others say college is the only way to be saved. What matters is listing companies you want to work for, reaching out and talking to people who make decisions, and finding out what the exact credentials are which you would need for them to hire you today. Those are the answers that matter, not random people's opinion on college. The two biggest things needed at small businesses and VC-backed technology companies are those who can learn really quickly, and those who are CEOs. The people who fight to control and master their emotions are those most valuable to small

businesses and VC-backed technology companies. These people have the attitudes which are highly valuable in small organizations. The right attitude at meritocratic companies is far, far more important than where you went to school.

Myth: It's not smart to join a company which might fail.
Truth: This is partially true. Yes, if you know a company is going to fail, then it may be stupid to join. But if the company can fail, but also has a chance to succeed in a big way—this is the sweet spot of learning, progress, opportunity, and risk. All these are hallmarks of innovation. What is so special about small business and venture-backed technology companies is the fact that they can fail. This puts us on the cutting edge of learning. This puts us in a place where we have stakes and incentives. Ultimately, being an entrepreneurial employee at companies of the right size ($1mm - $1 billion) is the first step to becoming a stoic operator in the face of risk. If the company fails, we'll be perfectly poised to move to the next small business or venture-backed technology company.

Myth: The goal is to never look foolish.
Truth: An unwillingness to look foolish is a character trait of failures.

Losers, Failures, and Actions for the Win

Losers and Failures: eat up all their willpower and mental ram each day worrying about looking foolish.

Winners: realize assholes exist, and that action, principles, ethics, and virtue will always appear "foolish" to somebody. Winners ask questions. Winners appear stupid to those who don't matter. They're willing to be thought foolish for extended periods of time. In fact, it's often those who can stand being thought foolish the longest that have the greatest successes.

Winners: Send cold emails. Make cold phone calls. Introduce

themselves to really successful people. They don't let fear of saying the wrong thing prevent them from making conversation. They don't care as much about their "elevator pitch" as much as they care about adding value. They ask for what they want. They hear "no" again and again, and aren't afraid to ask that last time when no turns to "yes." They hear yes and are careful to not press their luck, but also analyze when it's the right time to make a second request.

We shouldn't worry about sending emails and asking for things along the way while we connect. As long as we're genuine, we have nothing to lose. If you word something wrong, you'll find out quickly and be able to adjust. The people who are self-aware but willing to put themselves out there and risk looking foolish will be able to learn at a much faster rate than those who are unwilling to ever look foolish. The people who occasionally look foolish get a shot at everything they want, whereas those who care only about how they are viewed by others never will try.

Skyrocket your own learning by being willing to feel ignorant. Feeling ignorant just means you're being challenged at the appropriate level. It's not a sign to turn and run, or shut down in a conversation. Feeling ignorant is a signal to ask questions and listen, not shut down emotionally.

How does this view of not needing credentials for some of the most exciting careers and industries open up your realm of possibilities?

Here are some more ideas to build credentials, network, and create opportunities as you go:

- Set your Feedly with several feeds recommended by influencers in that space, or where influencers write online
- Download podcasts on your phone about the space you are interested in
- Examine the job openings at these places

- Reach out and establish contacts in these industries
- Contact sales people on these teams and talk to them (sales people will always answer their phone and email)
- Email industry contacts and executives and ask great questions
- Email industry contacts and executives and ask what skills they wished people applying for jobs had
- Write for publications in those industries and link back to your social profile
- Ask to work pro-bono on projects or help out at these companies, contingent on an interview down the road
- Comment in message boards and forums if you're too worried to become active and ask certain questions aloud

While discovering more ideas for exploration in these industries, unlocking opportunities and jobs is only a matter of:

- Finding what skills are necessary to break into these industries
- Finding out how to acquire these skills. You could look at LinkedIn profiles, and email people working in those positions and ask (they'll likely be incentivized to help by getting paid to bring in qualified candidates)
- Reach out to recruiters and make it known what you're looking for. Find out what skills the recruiters have seen consistently in demand, or how the recruiter will be incentivized to place you at a great venture-backed or mid-size technology company

On Reaching Out:

Ask questions, but try to tailor and orient everything to show you respect the value of the time of those you're reaching out to. Whenever possible, offer value first. Don't ask for anything difficult until you've confirmed you're delivering more value than you're being compensated for. Do enough research so it's easy for the person you're questioning to want to answer your question. Do enough research beforehand so you're not asking for what can be found online in five minutes. Make it easy for the person you're questioning to say yes. Appeal to their curiosity and their particular knowledge. Structure your questions in a way that acknowledges their experience and wisdom and communicates that you respect what they say.

On Following Up and Responding:

Show and demonstrate that you've made progress since the last time you spoke. Any way you can demonstrate the rate at which you're able to learn is incredibly positive.

Pro-Bono Work and Delivering Value First:

- Help them out (reach out after you've done your research to know exactly what that person needs that you can offer to do)
- Deliver value (do work that they actually need done)
- After you have delivered value, don't be afraid to ask serious questions such as, "Will you hire me for (insert job here), or will you give me a recommendation to someone who will? Will you recommend me on LinkedIn? If this type of work won't get me a job with you, what would?" (Get something in writing, and then deliver.)

Common Question about Pro-Bono Work:

What if they shaft me or take advantage of my free work, help, or the value I provide to them?

Don't waste time being angry or feeling defeated. If you're being taken advantage of, leave and cite your work as consulting as you approach the next opportunity. Add the consulting to your LinkedIn profile. Reach out to a competitor or place where the name of the organization you helped will be valued.

58 Virtuous Businesses Rebuilding America

"An ethical life isn't so when stripped of personal risks." –Nassim Taleb

"For no matter what we achieve, if we don't spend the vast majority of our time with people we love and respect, we cannot possibly have a great life. But if we spend the vast majority of our time with people we love and respect – people we really enjoy being on the bus with and who will never disappoint us – then we will almost certainly have a great life, no matter where the bus goes." –Jim Collins

Onto the Builders

Let's start with a quick word on these companies and what they need. We saw in the example of Google the type of individuals they are looking to hire. All of these companies, and venture backed startups in general, need people who can tolerate large degrees of uncertainty. They need people with courage, who are willing to take personal risks. They will pay a premium for those abilities, and we will have a chance to find ourselves at an employer creating value and re-making America.

Public Companies

These are some of the few publicly traded companies which still hold opportunity, or carry a powerful "I worked there" competency trigger or credential.

Amazon: A portfolio of businesses in varying sectors and of varying maturities. If you want to become involved with the future of retail (storefronts coming all over soon), and are obsessed with giving customers amazing service and options, this is the place to go. http://www.amazon.com

Apple: Steve Jobs was the grand master of sales for consumer technology. Apple will be the first trillion-dollar company. And, it's still early enough to become involved with Apple before they make and market smart everything with sensors. http://www.apple.com

Google: An amazing place to work. Opportunity abounds. If you want to start your own company, you have Google Ventures next door. Self-driving cars, various moonshot programs. Plus, the former CEO suggested laying fiber beaming Wi-Fi down to countries to change them instead of invading them.

Facebook: No explanation needed. Talented engineers are worth $500,000 - $1 million to Facebook. They just acquired Oculus Rift for $2 billion. Things have always been interesting at Facebook, and they're about to get fascinating.

Tesla: Gorgeous, safe, electric cars. Driverless soon. Elon Musk is an all around badass. http://www.teslamotors.com

Solar City: "At SolarCity, we believe in a better way to power homes and businesses at a lower cost while contributing to a cleaner planet."-[30]Elon Musk. Fossil fuels are great, but this is a business that's diversifying and making our species as a whole more antifragile. http://www.solarcity.com

Disney / Pixar: Steve Jobs got it going and developed it until it was acquired by Disney. Disney is doing amazing things, from starting their own universities to the usual innovations at their parks and the media they create. Overall, they're shifting and evolving the culture through amazing storytelling. http://www.pixar.com

Netflix: Streaming movies. Great work life balance. https://www.netflix.com

Tableau: Big data made simple and beautiful. Their yearly conference is a blast, and when I last checked, they had around 234 job openings. http://www.tableausoftware.com

Salesforce: Software changing everything, but things still need to be sold. Salesforce is a pioneer in the SaaS CRM space. There is still tremendous opportunity here. http://www.salesforce.com

GE: A massive, great bet for learning and stability. GE's Skills to Work program for Veterans is a great option for those interested in mechanics and machine technology. http://www.getskillstowork.org

Search and Local Search

Google is the incumbent, but there are two interesting companies about ready to crack search in different ways.

Duck Duck Go: A search engine for searches that remain private. https://duckduckgo.com

Foursquare: About ready to crack local, personalized, contextual search; if anyone does it, it will be Dennis Crowley at Foursquare. https://foursquare.com

3D Printing

Makerbot: Want to break in on the ground floor of 3D printing? I'm betting Makerbot will be one of the category winners and the first company to put a 3d printer in every home. This is the place to go. http://www.makerbot.com

Wearables

APX Labs: Founded by Craig Cummings and Brian Ballard, these veterans are serial entrepreneurs. They started and sold Ride Scout, and along the way built the technology they spun off into APX Labs. APX Labs is the future of wearables. http://www.apx-labs.com

Logistics, Infrastructure, and Energy

Nest: This company is making homes safer and saving huge amounts of energy. They were recently purchased by Google. Nest is bringing smart technology into the home, and the upside is tremendous. https://nest.com

Nextdoor: "The private social network for your neighborhood."[31] Building a community is crucial to bringing our country together. That's exactly what Nextdoor is doing. https://nextdoor.com

Uber: "Uber is evolving the way the world moves. By seamlessly connecting riders to drivers through our apps, we make cities more accessible, opening up more possibilities for riders and more business for drivers. From our founding in 2009 to our launches in over 200 cities today, Uber's rapidly expanding global presence continues to bring people and their cities closer."[32]

Uber has made traveling around cities better and more affordable than few imagined possible. They have raised a huge amount of venture funding but, I would argue, are seriously undervalued given the amount of global expansion they're throttling. The company has hundreds of open job listings as it is rebuilding and reallocating the resources of the world as we speak. https://www.uber.com

Shyp: "The easiest way to ship anything"[33] http://www.shyp.com

Oblong: The future of interacting with screens and visualization. http://www.oblong.com

Airbnb: "Rent unique places to stay from local hosts in 190 countries."[34] The sharing economy is the phrase that's used most often to characterize what AirBnb is up to, but they're doing so much more. Any company which allocates resources and taps into un- and under-utilized resources is a great thing. https://www.airbnb.com

The Future of Investing

AngelList: "Where investors meet startups"[35] Passionate about angel investing and private equity? Are you a great designer and/or coder who wants to learn under an A+ team? Definitely consider working for AngelList if you want to live in the future of private equity and master the concept of syndicates. Only a small fraction of accredited investors angel invest, and AngelList is perfectly positioned with the tools and network to capitalize on this massive opportunity. The founder, Naval Ravikant, is a highly skilled operator who has built massive companies before, and an angel investor in dozens of technology startups and businesses. https://angel.co

Daily Deals that are Actually Valuable

AppSumo: AppSumo sells discounted apps, technology products, courses, and generally all things that help make people more productive. Many people work for boring companies, with boring solutions to work related problems. It doesn't have to be that way! http://www.appsumo.com

Cloud and Security

BlueBox Security: With an estimated five billion Smartphones on the planet in the coming years, mobile data security seems like the right field, at the right time. https://bluebox.com

Box: Want to help enterprises and organizations move to the cloud? What about working with a boss who wears red tennis shoes and works longer and harder than just about anyone at the company? If you're a Chet Holmes disciple or a sales machine, Box would love to have you join the team.

The Private Investing Hub, Venture Firm, and Incubator in Northern Virginia

Disruption Corporation: A VC firm and incubator, Disruption Corp. is making an entire suite of products. http://www.disruption.vc

The Future of Education

AltSchool: "AltSchool prepares students for the future through personalized learning experiences within micro-school communities."[36] Want to fix K-12 education and inspire a generation of young people to follow their hearts and interests in a Montessori-style education like Sergey Brin, Larry Paige, and Palmer Lucky enjoyed? Great! AltSchool would be a great place to find the skills. https://www.altschool.com

There are many other companies reforming education that can be found on the portfolio pages of the top-tier investors we previously mentioned.

Health, Life Sciences, and Diagnostics

Between now and 2020, there is a projected shortage of doctors in the United States. Estimates range from 20,000-45,000.[37]

Predilytics.com: "Predilytics provides advanced healthcare analytic solutions to health plans, providers, and other risk-bearing entities. We transform healthcare decision making by revealing actionable insights on individual consumer health status and receptivity to engagement, and an organization's ability to realize impact."[38] http://www.predilytics.com

Doctor on Demand: "See a doctor from the comfort of your home, or on the go"[39] http://www.doctorondemand.com

Diagnostics (field): The diagnostics field is particularly interesting because in one swoop, the diagnostic tools which several companies are creating can instantly replace the need for tens of thousands of doctors and labs. To join some of the most cutting edge startups in the space, consider checking out the teams competing for the XPrize in Life Sciences.[40]

Payments, Currency, Cryptocurrencies, and Reimagining Finance

IEX: Just understanding how far finance and trading have gone from the old days of buying and selling a security is difficult. If you're ambitious about working on Wall Street, consider working for a company who is on the cutting edge of the future of trading in the public markets, the Investors Exchange. The IEX was started by Brad Katsuyama, the subject of Michael Lewis's amazing book, *Flash Boys*. If you're thinking about finance, consider IEX as an amazing starting point.[41] http://www.iextrading.com

Square: This is Jack Dorsey's second multi-billion dollar company. It's likely that square will be purchased by a tech giant soon. It's still a great time to get on board.[42] https://squareup.com

Coinbase: "Coinbase is a Bitcoin wallet and platform where merchants and consumers can transact with the new digital currency Bitcoin."[43] https://www.coinbase.com

Just about any company venture backed by a top-tier firm in the cryptocurrency space would be a good idea to join. Bitcoin might not be the long term winner, but there is a good chance it could be. Whether it's Bitcoin, an alternate version of Bitcoin, or another currency that owns the market, the Internet of money is just in its infancy stages. Most people have no idea how the hell it works. Layering skills is a great way to break into the cryptocurrency space. For instance, if you're an attorney, layering knowledge and becoming the first person to master the legalities around cryptocurrencies and Bitcoin would put you in the top 1% of experts in that space.

Kraken: Forex for Bitcoins and Crypto. https://www.kraken.com

Dwolla: "Dwolla is a payment network that allows any business or person to send, request and accept money. Their network securely connects to customer's bank accounts and allows everyone to send money quickly, safely and at the lowest cost possible."[44] This is an especially awesome opportunity for entrepreneurs who want to live in the Midwest. Dwolla has offices in Kansas and Des Moines, as well as on the West and East Coasts. They are tackling payments in a massive way. If you're interested in finance and payments, this is an amazing place to start. http://www.dwolla.com

Media, Journalism, and Culture

GoRuck: Helping fight the diseases of comfort in America through events, challenges, and gear. A budding lifestyle company. Veteran owned. http://www.goruck.com

Genius: "Genius breaks down text with line-by-line annotations, added and edited by anyone in the world. Through this crowd sourced annotation of music, news, literature, history, and just about any other text imaginable, Genius provides an interactive guide to human culture."[45] The web is stuffed with information which can be scary, overwhelming, and useless without context. Genius provides that layer of context. Plus, they were started by annotating rap music, which is what the young'uns want. Offering young kids knowledge at a place where they're naturally going for entertainment is a smart move, and this team has it covered. http://www.Genius.com

Inside: "Our Mission Is to Create the World's Best News Service." If you're an aspiring journalist who wants to go to a place with massive room for growth, led by a journalist-turned- entrepreneur, Jason

Calcanis, this is where you want to go. http://www.Inside.com

The Blaze: Love him or hate him, Glenn Beck's media company is bullish on America, and they're completely reimagining what it means to create, produce, and inspire Americans every single day. If you're an aspiring journalist, host, or passionate about American excellence and traditions, this is the place to go.
http://www.theblaze.com

Social Networks and Social Media

Path: The anti-social network social network. There is a chance Apple could have its eye on acquiring this startup as Path's founder, Dave Morin, was recently front row at an Apple event. https://www.path.com

Vayner Media: The premier social media agency in America for the Fortune 500. Led by Gary Vaynerchuk. http://www.vaynermedia.com

The Cuban Companies

Cuban Enterprises: Serial entrepreneur and billionaire investor Mark Cuban has a variety of portfolio companies doing amazing things in security technology. Any of his apps or companies is a great place to get involved with technology.
http://markcubancompanies.com

Consider these Cuban enterprises in particular:
CyberDust: http://cyberdust.com
Get Expire: http://www.getxpire.com

Politics

FiscalNote: Want to help fix politics? Instead of rolling in the mud, consider working for a company relentlessly innovating in the space and backed by top-tier investors. https://www.fiscalnote.com

Robotics, UAVs, and Drones

3D Robotics: "3D Robotics develops innovative, flexible and reliable personal drones and UAV technology for everyday exploration and business applications."[46] 3D Robotics is one of the major leaders in UAV and drones, and is doing amazing work in the space. The skies of America and the rest of the world will soon be filled (even more so) with drones of all sizes. They'll be used for thousands of different applications. Getting involved on the ground floor is an amazing idea. http://3drobotics.com

Modular Robotics: "Build your own robots that drive, crawl, slither, and glow. Elegant constructions with infinite possibilities!"[47] Robotics will catch on first as toys. If you're interested in breaking into this space, this is a great place to get started.
http://www.modrobotics.com

Airware: A platform of software for UAV's. "Airware enables customers to tailor drones to any commercial application by connecting aircraft, sensors, payloads, and application-specific software." http://airware.com

Big Data and Analytics

Gnip: "The World's Largest and Most Trusted Provider of Social

Data." http://gnip.com

Palantir: A big-data company which is preempting and fighting some of the biggest threats to humanity. They are allegedly responsible for finding Bin Laden. Want to fight terrorism with as little iatrogenics as possible? This is the place to do it. Started by Peter Thiel, this is the future of using data to preempt terrorist attacks. http://www.palantir.com

Adatao: Interactive big data. http://adatao.com

Space, Making Humanity Multi-Planetary

SpaceX: "SpaceX designs, manufactures and launches advanced rockets and spacecraft."[48] Once again, Elon Musk is crushing it in all aspects of business. Want to make humanity multi-planetary and save us all? You get the idea. http://www.spacex.com

Blue Origins: Jeff Bezos' space exploration company. http://www.blueorigin.com

Virgin Galactic: Richard Branson's space exploration company. http://www.virgingalactic.com

Stories of Reinvention Over Reintegration

To spark connections, curiosity, and insight into Veterans doing amazing things, below is a handful of Veterans working outside the traditional roles.

Tom Morkes is a West Point grad, Iraq War veteran, and even got paid to jump out of helicopters for a while. Tom is now an author, the founder of Insurgent Publishing and the co-founder of HighSpeedLowDrag.org - a business training platform for veterans.

"I recently launched HighSpeedLowDrag.org - a business training platform for veterans - with two incredible veterans and entrepreneurs - millionaire podcaster of entrepreneuronfire.com John Lee Dumas and men's style expert and founder of realmenrealstyle.com, Antonio Centeno. With a successful launch under our belt, we're now working on scaling and expanding the program to more veterans around the country.

Don't accept things as they are, nor put faith in the false believe that others will solve your problems for you. There's plenty of good intentions out there for vets, but the reality is: you're the only person capable of helping yourself. So surround yourself with great people, learn as much as you can, and take ownership over your own success. You'll thank yourself in a few years."

Danny Farr
CEO/Founder of Soldierfit/ Soldierfit Franchise LLC
President/Founder of 22 Needs a Face

He is a 13 year veteran of the US Army/Reserve. He lead the first team into the Pentagon on 9-11, and completed over 700 convoy missions as a .50 cal gunner from 05-06. He struggled with PTSD, ended up homeless, and attempted to take his own life.

"We are currently looking to open our third gym in Columbia MD, as well as launching our National movement by franchising our brand. We are also launching a Non-profit called "22 needs a face" that seeks to fight the epidemic of veteran suicides.

Getting help doesn't make you weak, and it is never to late to start over. I didn't know who or what I was 5 years ago. I had no life, I had no money, and I had no direction. I got the help I needed and since then I went from being a guy living in his jeep and feeling alone to a guy who, with the help of a wonderful team, has built a business that will gross $2 million this year and is well on its way to being a national company. If I was able to do this, then every single one of my bothers and sisters in arms can do it as well."

Ryan Micheletti is Director of Operations for Vet-Tech, the nation's leading startup accelerator for military veterans. Vet-Tech's mission is to transform the lives of veterans by empowering them to build companies. Ryan currently serves as a Squad Leader for a Search and Rescue Unit in the CA Guard and is the owner of Shield-Corps Security, a security company focused on hiring veterans.

"Over the past 4 years I have been building a national network of

technology executives, veteran entrepreneurs, startup advisors, venture investors, corporate partners and government entities to help Veterans become successful entrepreneurs. Vet-Tech recruits, trains, and funds veteran led companies and has a portfolio of over 40 companies nationwide. When I see the tenacity and ingenuity of the Veterans I work with, it motivates me go the extra mile. I really enjoy learning about their new technologies, from Solar to Body Armor, and helping them bring their ideas to light.

The most important message I have to share with Veterans is to start building your network as early as possible. We've all started at the bottom; attending meetups, going to seminars, building a reputation. It's easy to get discouraged in that paradigm - "This is too hard. I'll never make it. Why should I succeed where others have failed." However, the actions you take now will compound over time and set you up for major success in the future (Hint: It may take 4 years and we all know how fast 4 years can go by). It doesn't matter if you are looking for employment, want to be an entrepreneur, or going to school. Get out there and start meeting people that can help you get what you want out of life because you cannot do it alone. You are responsible for creating the life you want to have and the possibilities are unlimited. Find others that can help and you will be amazed at the "Net Work" that can be achieved."

Jay Knight: In 2008, former Sergeant Jay Knight left the Marine Corps to attend Southern Illinois University Edwardsville. After graduating college in 2013, he led a 4-person volunteer team in raising money for Child's Play Charity by bicycling across the country. They raised $14,700 in honor of fallen Marine Johnny Strong.

Currently Jay is working as the Operations Manager for a startup in Santa Monica called Porter. Porter is an in-home services app for on demand service needs such as housecleaning, dog walking, dry cleaning pick up and delivery. Some people have called Porter "the Uber of chores."

"There is no reason for you to not go to college. It is nowhere near as difficult as anything you have done in the service. It opens up a lot of doors for you career wise and you do learn important skills pertaining to your interests. Take internships, create clubs, and play sports. Your leadership, determination, and work ethic are more than enough for you to graduate with honors."

All of these guys are great examples of creating and taking your own path.

16

College Or Not?

"One of my friends has characterized the university system as the atheist church, which is sort of a successor to the Catholic Church…it's like the Catholic Church circa 1514…you have this priestly class of professors that doesn't do very much work, people are buying indulgences in the form of amassing enormous debt for the sort of the secular salvation that a diploma represents. And what I think is very similar to the 16th century is that the Reformation will come largely from outside." –Peter Thiel

"Institutions will try to preserve the problem to which they are the solution." –Clay Shirky

The Church of Education, Fundamentalism, and ROI

A polite disclaimer and warning: If you're a card caring member of the ***Church of Education****, I want to suggest you don't read the following chapter. I respect your right to choose your religion, practice, and worship freely. But you probably won't like this chapter, and that's all I have for a warning.*

It's a bit crazy that the average college student in America takes 5.6 years to complete a four-year bachelor's degree. On average, these students are left with around $30,000 of debt, and the ones who find jobs typically work outside their majors, in jobs they didn't need their degrees for in the first place. Collectively, there is $1.3 trillion of student debt! The indulgence of a college degree which doesn't open opportunities or job offers is a debt that can't be expunged via bankruptcy. Our government-run education system relentlessly sells students on blindly purchasing college degrees throughout grades K-12, but then wipes their hands clean when it comes time to measure the results. Luckily for them, the government controls and manages student loans, so they aren't incentivized to measure any results.

This isn't to say college degrees can't be extremely valuable; of course they can. But it would be insane to think that the groups who conventionally worship more "education" are producing great results for students. We've seen the type of companies producing results for America, now we need to find out how we can break into these industries. This requires double-checking, and objectively proving the value of a college degree BEFORE we buy it. It's easy to use the Post 9-11 GI Bill indiscriminately or carelessly. But if we've earned this benefit, it is much smarter to treat it with the same care as if we were spending cash.

In order to make the best college or not choice for ourselves, we first have to be willing to disobey the fundamentalist Church of Education (COE). The COE has largely developed over recent years into a branch of fundamentalists who don't want anyone to ever question the value of education. They teach that all other purchases are business and consumption decisions... but when it comes to *education*... well that's not something you can place a value on! How *dare* any of us question what education we purchase?! That's a quick overview of the fundamentalist mindset towards education. This fundamentalism has led to the $1.3 trillion in student debt, and it's one of the reasons why college costs (adjusted for inflation) have risen 400% since the 1980's. Our government gives out more grants for people to go to college, and universities smile and raise tuition proportionally.[49]

The first step towards disobeying the Church of Education means taking full agency over our salvation. We're not horrible sinners for questioning what we're paying for. Entrepreneurial leaders and founders don't accept dogma from existing institutions and culture; they create them.

We're allowed to scrutinize the business decision of purchasing a college degree. We are allowed to find places we want to work before we purchase a degree, and double-check the best route to land that position. We are free to double-check with those who have just graduated with a degree we're considering purchasing.

This chapter is about making the college or not choice in an objective, empirical, anti-fundamentalist way. There isn't a one-size fits all answer for everyone, and you won't go to hell if you find a path of salvation outside of traditional education.

If you're already an autodidact in an industry you love, or don't need any more traditional education, feel free to skip this chapter. But if you're curious about the recent debate regarding the value of higher education and are looking for some advice and strategies to make the college or not choice, feel free to stick around.

This chapter assumes that we have spent quite a bit of time learning about the industry and job we're interested in pursuing. Once we have identified our interests and struck up dialogue with decision makers in those industries (Chapter 15), we know a bit more about what we're getting into. We can ask the contacts if a degree will make a difference in getting hired, or if they will accept proof of our abilities in the form of a developer boot camp, design portfolio, sales numbers from a previous job, temporary work arrangement, or a GitHub profile.

We want to find the most affordable and best way to gain the specific skills we need to enter high growth fields with opportunity. We don't want to scrape by over the coming decades; instead, we want to make great money for doing great work. This chapter can also be useful if we find ourselves already enrolled in college and unsure if we're learning skills that command a high premium in the market. This is a tough choice, and sometimes it might make sense to stay and finish our degree, it might make sense to stay and supplement our degree through learning skills in other places, and other times it might make sense to set up other training we've proven leads to great jobs, and then leave college altogether. This is a tough question, and there is no one size fits all prescription. So in that pursuit, let's jump into the question of "college or not?" and discover if and when college is relevant to unlocking opportunity in modern America.

Discover if Others are Happy with Their Purchase

A few months before my last deployment in the military, I was thinking about going to graduate school. I wanted some sort of credential that could guarantee me a base level of income and opportunities. Looking back on where I was then feels a bit pathetic now. I was looking for confirmation from a university about the type of value I could provide to employers. I was looking for a "quick fix" of getting a graduate degree, prestige, and salary boost instead of doing the really hard work of making sure the degree I was going to purchase was valuable. At first I thought I'd go to law school and get a JD. I found the number of attorneys who had graduated and were clerking at $40,000 a year and changed my mind. Then I considered getting an MBA, but sadly I couldn't justify *why* I wanted to get one. I studied for the LSAT, GRE, and GMAT.

At this point I was waking up to the indoctrination I'd received from the Church of Education. I hated what I was studying. Then I got a bright idea… I already had a bachelor's degree in economics… I'd get a PhD in economics! That would be it! That's a credential that would set me up for success! As I began to study and take the prerequisites needed for a PhD in economics, such as calculus III, I started to wise up and be honest with myself. What the hell was I doing? Why was I doing something that made me miserable? Was it only because I had 30 months of Post 9-11 GI Bill that was sitting around? Not only was it boring, but there was no way this could have any value in the marketplace… could it? It took me this long to finally get the bright idea that I should just ask people who had purchased the degrees I was considering pursuing.

I decided I would reach out and discover the end result of a PhD in economics before I took the plunge into another four to five years of higher education. I had my sights set on a specific economics department at a specific school. So I found the dissertations and resumes of their recent graduates. I cold emailed all of them. A few responded, and we struck up some interesting conversations. I expected to hear that they were making great money, had learned a lot, or were generally so satisfied they'd made the PhD investment.

To say they were not satisfied was an understatement. Each person I reached out to seemed confused, unsure of the value they'd bought, and generally just wished they had studied more of what they wanted to. To be blunt, they all seemed to be harboring deep passive aggressive anger that had built up. They were all making the same amount of money they'd made before the PhD program. In a few cases, those recent graduates were still looking for jobs.

This was an icy wake up call about higher education. Looking to learn, and taking an interest in improving ourselves is great, but we should determine the actual potential value of what we're buying, or what our Post 9-11 GI Bill is purchasing beforehand. I talked to one of my former professors from undergraduate school for a final piece of advice before I decided whether or not to go to graduate school. He listened intently to my story of reaching out to recent graduates from the program. He responded with a few crystal clear points:

- Thank God you didn't waste five years of your life doing that
- Think about how much you could accomplish in the business world in five years

- Every book they read in class you can find used editions of on Amazon and read them on your own

His points hit home, and I got the wake up call I wish I had received before I devoted four years to getting a bachelor's degree. Making the choice to go to college or earn a credential doesn't have to be difficult. It only takes a small amount of extra work for us to "trust but verify" the value of what we're paying for. If we want to produce a certain outcome, earn a certain salary, or unlock certain opportunities, there has never been a more important time to trust no one, but reach out to discover what's going on right now in the marketplace.

On the Post 9-11 GI Bill

It's extremely hard to value "benefits" like the Post 9-11 GI Bill. It's not cash—can't be exchanged for cash— so it can be easy to use it carelessly instead of strategically.

College is many things to many people. It can be a place to learn how to think, a place to learn specific techniques and methods, a place to make important contacts, or to immerse ourselves in an atmosphere of new ideas. None of those things are impossible to do outside of college. What's difficult is to clarify what we want from college beforehand, and then do our due diligence to ensure we can't get what we're purchasing anywhere else. This is the case of most licenses and credentials (think nursing); they ensure us entry into high demand and high growth fields. These type of licenses and degrees are a no-brainer, but outside of them the harsh reality is that we can get all the reading material on our own. Then it just becomes a question of setting up stakes and incentives which will cause us to embrace learning.

It's easy to get the direct deposits of the Post 9-11 GI Bill set up and then forget we're making a purchase every single month. It's easy to just pick a college or major haphazardly and hope for the best. It's much more difficult to email recent graduates, ask for job placement data and starting salaries, and then go double-check that information with employers we actually want to work for.

The problem is, when most people talk about college, they are all talking about the average value of a college education. They're lumping together many people, many majors, many circumstances and historical time periods. Basically, they're arguing about something completely inconsequential to us. In economics, we learn that there is no such thing as an average cost of a joint product.

The statistics and figures that most news sources cite about college aren't helpful to an individual making a very important and very personal decision. It may be true that longitudinal studies over the past four decades shows that college graduates make more money than non-college graduates. But these snapshots of huge populations cloud and confuse the debate. If a political science major could not find a job after college and shifted into a field that did not need a degree but was very high paying (working for a rich uncle), he would be included in the "college educated" population, skewing the median salary of college gradates even higher. His salary ultimately gets included in the total data for those arguing in favor of a political science degree.

But the data that matters is what are college graduates making right now, and how does this compare to the graduates of developer boot camps (or whatever alternative we're considering here). The developer boot camp that costs $12,000, lasts three months, and ends with a starting salary of $80,000 at a technology company with our breakfast, lunch, and dinner included beats the hell out of most college degrees. Of course it's not for everyone, but the point is, it's easy to be deceived when we look at broad data showing the "average" value of college.

Finding our own truth about whether or not we should go to college is challenging. But consider starting here: ignore anyone who says, "Everyone should go to college" and avoid anyone who says, "Nobody should go to college." Neither view takes into account your individual unique situation, potential, talents, ambitions, personal agency, or existing skills.

The anti-fundamentalist truth about the Church of Education is that we don't need college to obtain an amazing standard of living. We're allowed to learn on our own, and if we're maniacal about our health, relationships, and crafting stakes and incentives around learning what matters to us – we can then think about learning outside of college. If we're unable to control and setup all those things, then college might be a great choice for us. The truth is we don't need a college degree to join many of the startups rebuilding America listed in the previous chapter. In order to join these startups and small businesses, we do need to be able to say:

- I have this skill you need (we can cite where and why they need it to showcase our knowledge of their industry).
- Here is a demonstrated, real world example of me using this skill and creating value (a consulting project, working app, testimonials from past bosses that showcase our abilities, freelance project, GitHub account, portfolio, etc).

We don't need college to become a multimillionaire. We don't need college to be a billionaire. But there are multimillionaires who have gone to college. They're the exception, rather then the rule. We don't need college to become a great person. We don't have to go to college because we have our Post 9-11 GI Bill collecting dust. But if we don't have any direction, ambition, or are patently lazy – then we need to hurry up and enroll to earn a degree or license where we'll be ensured a job. College can be useful to layer a skill on our existing skill set, or land a license or credential in a state-regulated industry where we see opportunity to learn about a field that interests us.

Just like the greatest essence of the American Dream is the

freedom to figure out and pursue our *own* dream, the same goes for college. The biggest thing we need to protect ourselves from is limiting our options. We don't want to be like so many young people in our country who are underemployed, or in debt. They trusted, but didn't verify. We don't have to trust, but anytime we pay money we'd better verify.

Now in the strategy and action section, we'll look over more ways and ideas to reverse engineer the answer to the college or not question.

Strategy and Action

"Top-down knowledge is an illusion. Education without erudition is nothing." –Nassim Taleb

"Thus the yeoman work in any science... is done by the experimentalist, who must keep the theoreticians honest." – Michio Kaku

Reverse Engineering Outcomes

1. Finding where we want to work.
2. Confirming that the area we're interested in is growing rapidly, or there is increasing demand.
3. Determine the skills needed for entry by asking employees, managers, and executives already working there.
4. If the skills and credentials they cite can only be obtained via college, consider the specific colleges that workers at the company you're interested in went to. The college might have a pipeline to that company, or vice versa, which will make getting hired much easier.
5. Reach out to the college in question to confirm placement rates, and get details on where those graduates are actually going to work (the person we contacted at the company in question might be a fluke or abnormality).
6. If the field we're going into DOES require a license or certificate to enter, we need to ensure we're obtaining that license or certificate in the quickest and most affordable way possible. We

also need to be sure the training institution or college has several real world examples of this license or certificate leading to great outcomes for students.

7. If the field we're going into doesn't require a license for entry, such as business, programming, technology, design, technical marketing (growth hacking, or just growth), then we need to figure out if college is the best path to entry, or if an alternative is a better choice. As of this writing, the most affordable and quickest way to land a great job in technology is through a boot camp. Developer boot camps are three to six month training programs largely focused around filling jobs startups need to hire for. The main types of boot camps revolve around training developers, designers, data scientists, and business developers. These programs are usually much less expensive than college, and most boot camps proudly showcase their placement rates and track records of helping their graduates find high paying jobs.
8. If we're still considering college, we can reach out to recent graduates from the programs we're considering applying to or are currently enrolled in. We can ask them: if they could do things over, would they make the same choice? We can ask about their salary (if they get offended this is just a helpful signal to ask questions from somebody else). We can ask how the college they graduated from helped them get a job. We can also ask them if they can help get us an interview right now. Generally, we can get the inside scoop on the real value of a degree before we make a purchase.
9. When in doubt, we can choose the path which gives us the most options: i.e. a degree with a credential or license, or a degree plus a developer/design/scientist/sales boot camp. Or we might find that although our Post 9-11 GI Bill doesn't cover a boot camp, we can use it for something like Computer Science, and raise money through a platform like Upstart to attend a boot camp over a summer, or start learning online and build a portfolio of work right

away. Just because the Church of Education might be weird, it doesn't mean we can't enjoy the occasional communion wafers and wine.

Here is an idea on reaching out with cold emails to help make the college or not choice:

Now, we can reach out to the employees of startups we want to work at or apply for. The quickest way to do this is to email them. A simple message could be based off a template such as:

Hi (Their Name Here),

I'm (introduce yourself, keep it short).

I'm hoping to get a job similar to yours in the near future. I'm excited about this industry and (their employer) because (whatever you can cite to demonstrate you know what you're talking about).

I'm curious to know what skills you have that helped you get hired. I'm thinking about (or already attending) (This college or traditional choice here, or considering this boot camp).

Which one do you think would give me the greatest chance of getting hired at (where they work)?

Thanks for your time!

Best,
(Your Name)

The tricky part about most job listings is that oftentimes the requirements listed aren't what's actually required. When we know the people making the hiring decisions, we're able to find jobs before they're listed, or build up enough rapport and demonstration of our abilities that we can bypass traditional job requirements.

Another idea to reach out after we've found a specific job of interest to us:

Take the company we want to reach out to and contact the CEO or other decision makers. Do NOT reach out to the hiring manager. Ask:

Hi (Their Name Here)

I'm a veteran and prospective applicant for (cite the job you want to apply for here whether it's open or not).

I'm hoping to get your expert opinion on whether pursuing Path A to get there (maybe that's a DevBootcamp) or Path B (maybe that's a BS in computer science) would be more valuable to your team at (their company).

I am hoping to get your opinion on the matter as I know hiring is time-consuming and expensive. Getting your opinion could probably save us both some time! :)

Best,

(Your name here)
(Social proof links to who you are)

Connectors and Training

"It is far better for a man to go wrong in freedom than to go right in chains." –Thomas Huxley

"If you can see your path laid out in front of you step by step, you know it's not your path. Your own path you make with every step you take. That's why it's your path." –Joseph Campbell

Opportunity and Connections Everywhere

Now let's jump into some of the best connectors and training. Many of these platforms can be utilized to help discover if a college degree, boot camp, certificate, or some other credential is the best route for us and our dollars.

LinkedIn

This is by far the most underrated social network for gaining intelligence on what type of employees businesses need. It's also an amazing tool for identifying the exact training needed to obtain jobs. Almost every university is set up on the platform. We can get on LinkedIn and view the recent graduates of these universities, see where they're working, and message them. We can also reach out to these universities' admissions offices and get answers to our questions. From LinkedIn profiles, we can study recent graduates of the undergraduate or graduate degree we're considering. From there, we can study which majors lead to which jobs, and identify the quickest path into a job we want.

If You Need a Job ASAP

Hire Purpose: Helps veterans find a career they'll love, with employers who would love to hire them. https://www.hirepurpose.com

Vets in Tech: "VetsinTech supports our current and returning veterans with re-integration services, and by connecting them to the national technology ecosystem." Want to break into technology but still have no idea where to start? http://www.vetsintech.co/about/

Training in Growing Technical and Manufacturing Fields

Warriors4Wireless: A Virginia State University program designed to train veterans to get credentialed as wireless technicians. http://warriors4wireless.com

Almost every large community college in every state has a path through their school to a high-demand job which can only be accessed via license or credential. These are definitely worth considering if a traditional four-year academic degree or boot camp doesn't interest you.

Accelerated Entry into a New Field with High Growth

Skills to Work: Skilled technical manufacturing is a rare skill set these days. Skills to Work is a program run by General Electric which trains and hires veterans for future-proof jobs in technical manufacturing. Robotics will certainly dominate the future, but for now they need to be operated and understood by skilled technical manufacturers. If you want to work with your hands instead of only with technology, this is an amazing opportunity. getskillstowork.org

Mike Rowe Works Scholarship: Mike Rowe is spearheading an initiative offering scholarships for those who want to learn trades which are in high demand. http://profoundlydisconnected.com/foundation/

Upstart: Started by an ex-Googler, this is the platform you want to go to if you need funding for any venture you're working on. If you need to raise money for attending a boot camp not covered by the Post 9-11 GI Bill, Upstart is a great place to do it. http://www.upstart.com

Learn How to Code, Design, Sell, or Market in Person

The Iron Yard: If you're on the East Coast or down South, the Iron Yard is a must to consider. http://theironyard.com

Starter School (Chicago): A company which is part of Basecamp (formerly 37 Signals), I trust them to do boot camps the right way. http://www.starterschool.com

Starter League: Similar to Starter School. http://www.starterleague.com

Jumpstart Labs: A boot camp and training company that "builds developers." http://jumpstartlab.com

DevBootcamp: The original developer Bootcamp. A fantastic choice to accelerate your learning and break into technology. http://devbootcamp.com

General Assembly: One of the premiere locations to learn skills necessary to break into technology fields. Locations all over the country. https://generalassemb.ly

Galvanize: CO, CA: http://www.galvanize.it/school

Full Stack Academy of Code: NY, NY. http://www.fullstackacademy.com

Tradecraft: "We train smart people to succeed in traction roles at high-growth startups."[50] http://tradecrafted.com

Turing School of Software & Design: Denver, CO. http://turing.io

Learn How to Code, Design, Sell, or Market Online

Bloc Web & App Development, Plus UX Design: https://www.bloc.io

Viking Code School: http://www.vikingcodeschool.com

TeaLeaf Academy, GA: http://www.gotealeaf.com

Exercism is a site updated daily with new programming tests and challenges.http://exercism.io

CodeAcademy: Learn to code for free, right away. http://www.codecademy.com

Thinkful: Learn Swift, the updated programming language for iOS apps. http://www.thinkful.com

Treehouse: "Learn how to build websites & apps, write code or start a business."[51] Want to start learning web design or coding for the 21st century jobs? http://teamtreehouse.com

Accelerators, Incubators

Expa SF, NY: "Design + Data"[52] Led by Garrett Camp, this is part accelerator, incubator, and VC fund for early stage companies. http://expa.com

Techstars Patriot Bootcamp: A branch of Techstars devoted specifically to veterans. http://www.techstars.com/patriotbootcamp/

Techstars: If you don't need the soft start of Patriot Bootcamp and have users, revenues, advertising dollars, or growth, go big at Techstars. http://www.techstars.com

YCombinator: The top accelerator for startups worldwide. They've funded over 700 startups, several of which have gone on to become multi-billion dollar businesses. http://www.ycombinator.com

FishBowl: An incubator for technology companies which is free to join. They're based in northern VA at the AOL campuses. http://www.fishbowllabs.com

DreamIT: Baltimore, DC, and Austin http://www.dreamitventures.com

500 Startups: Focusing on funding mid-size startups in CA and expanding worldwide. http://500.co

AngelPad: LA, http://angelpad.org

Co-Working Spaces and Incubators

WeWork: Both an accelerator and incubator space, quickly expanding nationally. https://www.wework.com

1776: Evan Burfield's startup incubator. "The world's biggest challenges. 1776 is passionate about reinventing our lives as citizens. We're creating a global community by connecting startups with the resources they need to grow and succeed —from mentorship and corporate connections to capital and media attention—in sectors such as education, energy, health and cities."[53] http://1776dc.com

Disruption Corporation: Paul Singh's Northern Virginia co-working incubator company provides access to his crystal city Tech Fund, making this a very compelling place for those building the future in Northern Virginia.

Vet Tech: This is a high-quality accelerator and incubator for veteran-owned and led businesses. They're quietly amassing a large portfolio of veteran-owned and led companies. http://vet-tech.us

Funding for a Business

There aren't any wrong answers to fund a startup or business pursuit. There are only funding options which leave you with fewer, or more, options. Some options to fund a startup which leave you with fewer options might be a loan, or venture capital investment. Funding and growing a business with cash and revenue typically leaves us with more options. At a certain point in a startup's trajectory, investment from the right investors makes perfect sense. The best places to get started learning about startups that are growing, as well as how they're funded is:

Angel List: Angellist.co

AngelList helps match startups with the best accredited investors and syndicates, many of which provide much more than just money. If all you're looking for from investors is money, your business will probably not succeed. Investors or angels from the trenches of the industry in which you're starting a startup have connections and experiences that are especially valuable.

If your startup is already pulling in huge revenue, or you uncovered a steady stream of secrets that is enabling you to build a monopoly 10 times better than any competition, then great! Angel List lets you use your numbers to attract great investors to you. Or, you can pitch the top-tier VCs below, which are the firms rebuilding America at scale.

Top Tier Venture Capital Firms

A16Z: Want to work at the hub of an amazing venture capital company? Getting hired or funded by A16Z might be challenging, but this firm is the best of the best. http://www.A16z.com

Lower Case: This is a venture firm started by Chris Sacca who came back from over $2.1 million dollars of personal debt… in his own name! Rather than file for bankruptcy, Sacca paid back every dime of his debts, and wrote credit card check investments into some of his first major wins as an investor. He invested early in Twitter and Uber, and now has a firm managing well over $1 billion. http://lowercasecapital.com

Founders Fund: If you've read and resonate with Peter Theil's amazing book, *Zero to One*… this is where you want to go. Zero to one innovations future-proof humanity, whereas spreading 1 to n put us all at risk. http://www.foundersfund.com

Foundry Group: Brad Feld's venture fund, they also run Techstars and Techstars Patriot Bootcamp: http://www.foundrygroup.com

Google Ventures: The venture arm of Google with a roster of 250+ companies, many going after huge "moonshots." http://www.gv.com

First Round Capital: A firm with a great reputation for treating founders with respect and an impressive portfolio of companies. http://firstroundcapital.com

Upfront Ventures: A founder friendly LA venture fund. http://upfront.com

True Ventures: "We are serial entrepreneurs ourselves: the True partners have founded 14 startups, and three of our partners are active Founders and CEOs of companies in today's technology ecosystem."[54] http://www.trueventures.com

Part 5: Entrepreneurship

17

Return of the Polymath and The Passion Myth

"A human being should be able to change a diaper, plan an invasion, butcher a hog, conn a ship, design a building, write a sonnet, balance accounts, build a wall, set a bone, comfort the dying, take orders, give orders, cooperate, act alone, solve equations, analyze a new problem, pitch manure, program a computer, cook a tasty meal, fight efficiently, die gallantly. Specialization is for insects." –Robert Heinlein

"Nature loves courage. You make the commitment and nature will respond to that commitment by removing impossible obstacles. Dream the impossible dream and the world will not grind you under, it will lift you up. This is the trick. This is what all these teachers and philosophers who really counted, who really touched the alchemical gold, this is what they understood." – Terence McKenna

Skills Always Pay the Bills

We're living in a time period where theoretically, it's never been easier to become a polymath. But in practicality, there can be many cultural and social hurdles to this achievement. The idea of becoming expert at many things seems daunting to many people, so they never begin. As time elapses, it's easy to fall prone to the sunk cost fallacy. This is a type of illogical reasoning where we resist changing course because of all the time or resources we've already wasted or invested on a particular path. This is foolish because the earliest time we have to begin anything is in the present moment.

So why is it difficult to become a polymath or monopoly of one today? If we aren't careful about our relationships and assert our intentions and ambitions to become a polymath, or embark on some huge undertaking, we'll likely meet warnings to "slow down" or "just focus on one thing first." This can sometimes be prudent advice. But oftentimes the people who voice their concerns about ambition have bought into the lie that we must either fully associate ourselves with our career, or keep our work completely separate from ourselves and our identity. There are no rules around the amount of interests, pursuits, or skills we develop. It's perfectly fine if we have trouble explaining to empty suits what we do. Being too complex to define is a noble pursuit. Having many interests, projects, and ventures helps us buffer ourselves from boring people and helps bring interesting people into our lives. Those who have ambitious, virtuous intentions are most likely to recognize each other. This is important to remember as we go about acquiring our skills.

The secret all polymaths know is that it becomes easier to learn when we increase our workload. When we have many things going at once we can cycle through projects appropriately. In order to tap into the full power of our brains it helps to have many projects, sometimes dozens, going at once. Another particularly helpful way to supercharge our brain is to develop goals and ambitions so large they spur us towards massive action. These are the only mindsets which make it possible to learn quickly, layer our skills, and become a master at what we aspire to. We receive our minds, the most powerful computers in the world, free at birth. This leads most of us to never value them appropriately. We only begin to scratch the surface of our potential when we build huge ambitions, then involve our minds in the daily grind of building habits, tackling many projects with deadlines, stakes, and incentives. This is the only place where we'll accelerate our ability to learn.

For those who have the courage to experiment with this process of ambition and strategically layering skills, we'll find our capacities increasing. It might be a challenging time socially and culturally to begin to aspire or learn many things at once, but once we get started, we'll be glad we did. We can't help but thrive economically, find ourselves or our passions, and become a monopoly of one along this path . In fact, if we aim to access and create the type of value and innovation which will rebuild America, we must begin now.

There have never been more low-cost learning resources available to speed up our learning. Just consider the humble podcast for instance. This is something many take for granted, yet the technology behind it is extraordinary. On any given day, online or directly on our phones, we can listen to the best teachers in the world, from any given industry or field of study. This morning I listened to two separate interviews where a billionaire answered a plethora of questions from the show host. He wasn't just answering questions; he was distilling wisdom that's potentially life changing if it's applied. We can't personally hire someone like that for coaching, but we can listen for next to nothing thanks to a podcast. Podcasts are an incredibly undervalued way to receive digital mentorship from the best in the world. For those who have the faith, willpower, and patience to teach themselves, the time for erudition is now.

The standout men and women of history have all been self-taught, multi-skilled, life-long learners. Leonardo DaVinci, Lucius Seneca, Ben Franklin, Carnegie, Jobs, Musk, Cuban, Rowling, Branson—or whoever we want to consider a modern-day polymath—started far behind where we are in terms of resources and technology.

These men suffered through adversities and were able to prosper without all the advantages we have in front of us today. They managed to become self-taught masters of many skills. No matter what circumstances we find ourselves in right now, others have triumphed over worse.

But how the hell did these guys—the ones who didn't come from comfortable situations—do it? The simple answer, which we covered earlier and that most ignore (even after hundreds of years of it being repeated!), is they started learning through doing. They read, they wrote, they made sure they were around others doing the same. They came up with ideas and then figured out how to begin pursuing them immediately. They started small projects they tinkered on; they allowed their creativity to guide them from one project to the next. They didn't look for their passions; they looked where it was easiest for them to invest huge amounts of effort and willpower.

We need to begin rotating through many small projects in a strategic fashion. We can see for ourselves where we work the hardest, and when it makes sense, we'll know when to double down and go all in on our pursuits. People who are self-taught don't judge themselves for leaving projects unfinished. Leaving many things undone is incredibly important. When we have multiple projects going, we develop the important ability to quit when it makes sense. How many individuals have we seen on Shark Tank who are pitching ideas they should have quit years ago? Developing the ability to quit at the right times, without ridiculing ourselves, is incredibly important.

This follows the natural ebb and flow of our interests. Our desires and passions don't define us for life unless we insist they do. Let your situations, interests, and pursuits be incredibly volatile. We need to become comfortable juggling a large number of projects, discarding foolish pursuits, then capitalizing and dropping everything for months at a time when an opportunity presents itself. The loser looks for work life balance. The winner looks for where it makes sense to go all in.

One of my favorite cartoon strips of all time is Calvin and Hobbes. The creator of the cartoon strip, Bill Watterson, was a smart guy but he rarely gave interviews. During one of those rare interviews he was asked about creativity, productivity, and recharging. Watterson said,

> *"Shutting off the thought process is not rejuvenating; the mind is like a car battery—it recharges by running."*

This phrase can feel overwhelming until we stumble over the power of it for ourselves. When we experience it after a ten hour day of work has left us drained, and we switch over to learning that new thing, then Watterson's words ring true.

So let's begin the practice of starting a campaign of projects in the pursuit of acquiring skills. One of the last hurdles to beginning to learn many skills at once is that most people think we have to become world class at something before it gets interesting. They severely overestimate how long it takes to become skilled enough at new pursuits. This is unfortunate because there are many things we can become "expert enough" at within only a few months. Expert enough, meaning we can even be paid for the work we create. As we saw in previous examples, in only three months time, developer boot camps around the country are aggressively placing graduates at jobs paying close to six figures.

The process of time necessary for skill acquisition on the road to mastery has been discussed extensively by several authors and researchers. These notable authors include Malcolm Gladwell, Timothy Ferriss, and Robert Greene.

Gladwell says it takes around 10 years to become world class. Greene takes a more strategic, Machiavellian-meets-Aristotelian view and says 10-20 years. But Ferriss comes in with the exciting reality that if we use the right stakes and incentives, we can become expert enough in a few months.

Mastery: top .01% of a field - takes 20 years

Budding Master: top 1% of a field - takes 10 years (outliers)
World Class: top 3% in a field - takes 5 years
Expert: top 5% in a field - takes 1-3 years
Expert Enough: top 10% in a field - takes 3 months to 1 year

More authors and researchers are beginning to emerge showcasing that expert enough can be accessible after only a few months of highly concentrated effort, stakes, and incentives. In his excellent book *SmartCuts,* author Shane Snow advocates studying the outliers, or "fat tail" top performers in order to uncover the most sustainable paths that are both short and smart. Our culture has likely left us with warning alarms going off in our head at encountering advice like this… shortcuts? Taking the easy route? If our intentions are in the right place, and we're seeking good things, why wouldn't we take every smartcut we could? From the Presidents of the United States, to CEOs of technology companies, it's often the youngest and lesser experienced that don't know the "way things have been done" who figure out "the way things SHOULD be done" and then do them.

We're also prone to underestimate the small victories we'll unlock along the way towards "expert enough." And those small victories only compound. Many people never begin to learn because they severely underestimate how much joy lies in getting to each of these levels of expertise. At the expert-enough level there are plenty of ways for us to be compensated for our newfound skill, even compensated really well. Also, consider how we can make the math part of the mastery equation work in our favor. To become the top . 01% in a given field, it may take 20 years. But to become the top . 01% in a brand new field by the combination of skills we've layered to create? Now we're talking only a few years.

Think about the person who chooses a few skills which are in high demand. They become expert enough at them, reap the rewards, and then repeat the process again and again. They might be one to three years into their learning, but they are not competing on the same playing field as the person who has focused on only one thing, sunk in eight years of agonizing investment, and is wondering why he's gone nowhere.

To become a monopoly of one and create a fulfilling and opportunity-packed life, we only need to concentrate on becoming expert enough at that first small high-demand skill, break it down to a pursuit that only takes a few months, and then layer it with the next high-demand skill. This path makes the most sense logically and competition-wise. Why on earth would we want to spend 20 years devoted to becoming a master of a field overwhelmed with competition? It would be nearly impossible to become the leading computer scientist in the world without putting in at least 20 manically-focused years. But if we want to get started on becoming a world class polymath in technology, we could start learning the brand new iOS language, Swift. In only a few months, we'll have a skill people will pay us for. Then we could layer on the ability to write in Swift and port apps to Android. Instantly, we unlock more options to work at more technology companies, to freelance, or choose a VC-backed startup. Later, we can layer on JavaScript and other languages and competencies. In a short period of time, the person who does this will become a world class full stack-developer. The person who takes the 20-year path to expert computer scientist status will likely be eating ramen noodles as a teaching assistant while we're busy building the future. Now let's look at some of the premiere mindsets to help capitalize on opportunities around us.

The Premiere 21st Century Mindset, Goal, and Path

Some people hate the word "goal." I agree that sometimes the definition can be murky. If we only focus and work towards one goal, we can find ourselves in a perpetual state of discontent. We tend to grow frustrated because we haven't achieved it yet, then once we do, we become bewildered when the initial euphoria of victory passes.

This is why a diversified set of goals, interests, and even massive future projects are so important. With a basket of smaller goals, we can break each down until they are a series of little steps, with little victories, in order to consistently achieve the euphoria of, "winning." Once we begin spreading and breaking down our goals into manageable chunks, we can then pick out hyperspecific skills to begin learning. There are many small hyperspecific skills we can teach ourselves on the way to achieving broad and overarching goals. By starting with a massive goal, skill set, or pursuit, we can then get busy breaking it down into small, localized "goals" that we can begin learning. From there, we can learn the next skill, then the next, until we gradually check off all the small goals necessary to achieve our big goals. On a path of learning like this, we'll eventually find ourselves with an immensely valuable and unique set of skills. We'll likely be at a place where competition is scarce. If we keep up the practice long enough, we'll find ourselves as a monopoly of one.

The Mindset Goal

Apatheia: The mental state where nothing can be taken away from us. Impossible? Most would say yes. But we're not most people.

The Skills Goal

Autodidact: A self-taught person who learns best through self-teaching. Once we can teach ourselves, we can move to becoming a monopoly of one. Impossible? Not at all. Reading. Podcasts. Taking risks.

The Only Path We Need to Worry About

Identify the skills we can learn which employers are ready to pay a premium for immediately. Learn those skills and take care of our economics and finances. Use that first skill or skill set as an economic launch pad to acquire more. Layer the next skill into our repertoire until we become a monopoly of one.

A person who pursues the mindset of apatheia and undertakes the actions of an autodidact will eventually become antifragile. They will also, inevitably, become a monopoly. This means they will have built up skill sets unique to their interests which by nature of the rare combinations, make them roughly ten times better at what they do than anyone else.

These are some of the most important mindsets and ideals of the men and women who will build a future that is free, prosperous, and one in which humanity survives and thrives.

Strategy and Action

...Every business [or for our purposes, individual] is successful exactly to the extent that it does something others cannot. Monopoly is the condition of every successful business." –Peter Thiel

"I'm not trying to find nobody else to beat." –Drake

Skill Layering Until We're a Monopoly of One

If we want to become a world class designer, we might first practice sketching, then Photoshop, then take on clients or book covers, or designing UX and UI for apps. As we learn each skill and offer it to clients or the market, we discover where the biggest margin lies. Where the biggest margin lies is where there is an inefficiency in the market. The biggest margin where we're delivering value in an ethical way is the most virtuous place we can apply our skills.

If we want to become a world-class speaker, we could start offering gigs on Fiverr, oDesk, Elance, or ACX for voiceover work or narration. We could then find some speaking gigs, start for free, then take paid work, then maybe take a sales job that forces us to speak articulately at length. Then we can move to speaking at conferences for audiences (at first for free). From there, we can eventually charge for our services.

By now, you've heard me beat this idea into the ground. Competing in places where there is fierce competition for small rewards is the laziest waste of talent imaginable. The truly virtuous seek out and serve areas of the market where there are high margins, or where they can apply their skills to serve as many people as possible at once (when low margins are justifiable). Monopolies don't signal "lack of fairness"; they signal someone or a group of people providing value to others in a truly innovative way.

There is a classic example of layering skills I bring up when I talk to groups of people, especially students. Many of them want to break into technology, yet they're just getting their computer science degree. That's a great degree to have, but they must become more proactive in acquiring and layering skills to build the monopoly of themselves.

Example of a Skill Layering Path in a High Demand Field

1. Learn the brand new Apple iOS Swift programming language.

All we need to know about Swift is its way easier to learn than its predecessor, Objective C—it's a great time to get in! Apple will be the first trillion dollar company, so it might be a good idea to learn the languages they're building.

Existing Swift developers have said the language is a huge simplification which creates an amazing entry point for anyone who wants to get started in the industry. If we learn Swift, with a bit of Xcode, we immediately have a skill people will offer us money for. At this point, we know we're literally helping make the United States economy more efficient. We can get started with a free book Apple has published detailing the Swift language. There are also online classes focused on Swift available via Thinkful.

From this point, we could learn Photoshop in order to do a little front-end work (if we are design inclined) or we could begin learning more about back-end work and APIs. This is just my non-technical explanation and brief skill layering overview, but you get the idea.

2. The next skills to layer onto of the existing competency of Swift

would be perfecting Xcode and Cocoa

3. From there, JavaScript can be extremely useful for learning the back-end of the apps we're building.

4. Next, we can learn Amazon Web Services to become completely deadly and either start our own business or work to build apps for others. But at this point, there is a much better chance that we've received offers we can't refuse and are being paid $150,000-$250,000 a year at a Silicon Valley technology company (and having breakfast, lunch, dinner and childcare paid for while we work).

5. But want to layer more skills? We can keep going and learn how to efficiently convert the iOS apps we build to Android via a tool like AppFolio. Now we've moved even further toward creating a monopoly of ourselves.

6. From there, we go deeper into learning the existing skills we have. By this point, we'll have found several niches with great margin and unique opportunity. Maybe we see an opportunity to layer on skills that involve learning sensors, HUDs, and wearable technology; the opportunities are endless.

The exercise below is to search for existing opportunities to start layering skills. Start below by writing down a skill people will pay for, then practice writing out skill layering progressions. How will we progress so with each step the air becomes more rarified, and we compete less and less?

We can even get technical and look up specific skills to estimate how many other people are proficient with them, and then study future projections and demand for them.

1.

2.
3.
4.
5.
6.

Some additional questions we can ask ourselves when going through this exercise:

With what skills are we already among the top 5-10% in the world? How are we estimating this? How could we use data to better estimate this? As a member of an all volunteer military, we're likely in the top 1-5% of Americans in the courage department.

What brand new recipes of skills could we create? For instance, if adding a Dev boot camp to our college degree or IT experience would only take three months, that would be the first ingredient in a great recipe.

If we've become proficient at teaching a specific type of class or skill in the military, how can that directly translate to the private sector? Is there a simple skill in the private sector we could layer onto our military training to compete even less with others?

For instance, if we were a diver in the military, could we layer a few civilian diver certifications onto that training? Then, what if we layered another even more in-demand credential like underwater welding? It's not long before our work commands a premium.

We've already done a hypothetical layering skills path above. Now let's consider another where we capitalize on where we're already at. We can consider looking extensively at places where we have an unfair advantage of practice. Maybe we are amazing speakers from giving so many briefings or spending so much time on the phone while we were deployed. Maybe we have an analytical, left brain mind, with an obsessive attention to detail, in which case programming could be an amazing fit for us. Maybe we've worked and built things with our hands and are fascinated by building and machining.

Below, try the sample skill layering practice again by starting with where we already have invested time or have expertise. Feel free to continue to go through this practice. When our minds continually think in terms of finding less competition, our service to our families (and all of humanity) continues to increase proportionally.

1.
2.
3.
4.
5.
6.

18

Survive, Grow, Signal

"The art of survival is a story that never ends" - Christian Bale, American Hustle

All courses of action are risky, so prudence is not in avoiding danger, but calculating risk and acting decisively. Make mistakes of ambition and not mistakes of sloth. Develop the strength to do bold things, not the strength to suffer." –Niccolo Machiavelli

The Path of the Founder, In Three Words

The American media is fascinated with what the end result of entrepreneurship looks like. But the actual path and beginnings of every great entrepreneur is rarely discussed. In our society and on the news, we often don't realize just how many times we're looking at only the end results of years of effort. The media often creates a short bite at a glance studying a "success." The person interviewed has only seconds, or at best minutes, to speak. The "successful" person then has to summarize years of effort, stopping, starting, adjusting course, faith, and courage into a few sentences. This inevitably leads to them having to try to distill years of specific actions into a sound bite. Since birth, we've been conditioned (often unknowingly) to view the end result of success with an explanation from the alchemist which sounds as though their success was attributed to genes, luck, randomness, or serendipity. The most useful definition of success I've ever heard is based on Earl Nightingale's masterful definition. He stated that success is "the progressive realization of a worthy ideal." This definition of success carries with it some hard truths. Those truths include the fact that *success* is available to any one of us right now. We can each put our faith in beginning to pursue a worthy ideal. We don't even need to have a concrete ideal. As long as we're seeking worthiness, we can figure out the rest along the way. This is why when people who are truly successful talk about how they got there, it's shunned by most traditional media. When Twitter and Square founder Jack Dorsey tweeted in 2013, "Success is never accidental," it angered the thousands of media pundits, trolls, and hurting people who had bought into the mainstream media's manufactured indoctrination of success with luck. If we can push past any initial previous indoctrination, and in some cases brainwashing, about *success*, then we can immediately get busy being *successful*.

The truly successful person sets off on the journey, becomes charged with good intentions, and puts faith in the truth that they can learn everything they need along the way. They know many of their pursuits, bets, projects, and ventures will take years to bear fruit. They also don't let this stop them from getting started today. If we were to distill these years of effort toward success, of becoming more entrepreneurial and eventually becoming a founder into just three words, they might be:

Survive, Grow, and Signal

The successful person first masters the art of survival. This includes mastering their diet, sleep, and exercise. Then they become CEOs (the emotionally aware kind). They build the powerful social and mental competency triggers through reading, writing, and speaking. They enter the market and learn skills in great demand. They layer more skills onto this practice and develop multiple streams of income. At this stage, the budding success (future founder) is mastering the art of survival. They cannot be killed because they're properly valuing everything as they go. They have their finances mastered and enough money coming in to fund their growth. Mastering this art of survival is incredibly challenging, and it's the stage where most people give up. Once we master the art of survival, we become a net positive to every interaction we have with others who are seeking to be a success.

They don't realize that after mastering the art of survival, life becomes amazing; relationships blossom, opportunities present themselves, and mentors appear. At this stage, growth and progress have become part of everyday life. They continue to grow and progress further. Then, they signal to the right people who can join them in their growth, which means joining them in their pursuit of a worthy ideal. This might be a spouse, a co-founder, an investor, employees, friends, job offers at a business that is changing the world; the list goes on. The successful person masters the path of progress and learns to: Survive, Grow, and Signal.

Survive

Our first goal on the pursuit to success is to survive and continue to increase our exposure to upside. We survive, tinker, experiment, reach out, and continually take small risks which can't kill us while pursuing our opportunities. To illustrate and start this process, we need a funnel to capture opportunities while we research just what it is we want to do and learn; and what business we want to work for. Uncapped upside means giving ourselves the opportunity to join a startup where we can earn equity instead of salary. There are those who become salary and perk snobs, and there are those hell bent on getting *the right* few points of equity. Those seeking equity are seeking to own the process of making things better. They're investing in the long term with their actions. Those seeking more pay and perks are generally of the looter mindset.

The large part of mastering survival is through starting side businesses and developing multiple streams of income. This gives us the ability to not be devoted to a single source. Part of survival means recognizing that fate will bring us difficulties. By mastering our finances and creating several sources of income, as well as a network of people we've helped and who trust us, we become immune to economic fate.

Mastering the art of survival means making a commitment to ourselves and our families to thrive—no matter what. Whatever life brings us, we will use, mold, and force it to serve us. When we do this, we create power over our lives and situations.

When we commit to thrive no matter what, we begin to become antifragile. Time guarantees volatility. The person who masters the art of survival begins to love time, because volatility brings opportunity and new information. Of course it's one thing to verbalize a love of fate and volatility; it's quite another to get laid off and send out 500 emails in a row trying to create value and tap opportunities in our existing network. This mentality involves always extracting a lesson from life no matter how shitty things become. Nietzsche and Taleb have created amazing bodies of work revolving around how to love and capitalize on fate.

Nietzsche famously said, "Amor Fati" - love fate.

We might remember Taleb from his presentation of the word that sums up the interesting half of life, *Antifragile*. It's Taleb who helped translate the more aggressive branch of Roman Stoicism into the modern vernacular. Taleb articulates the mindset of the modern stoic as one who says, "*Fuck you to fate*." Saying "*fuck you*" to fate and any bullshit brought to us by volatility (time) is an incredibly healthy and proactive mindset. A loud *fuck you* to fate declares and commits us to figure out how to alchemize the uncertainty which fate brings.

We are not the spineless victims of fate. In Breaking Bad parlance, "*I am not in danger, I am the danger... I am the one who knocks.*" This is the attitude which primes our mental faculties to capitalize on what fate brings. Once we have mastered the art of survival and flip off fate as we go, the next imperative becomes to *GROW*. Without continual growth and progress it's impossible to ever be truly satisfied with life.

Growth

"Survival is a growth strategy." - Ian Hogarth

Once we've mastered survival, we naturally set the stage to master growth. When our pursuits can't be destroyed by the kinds of maladies which usually kill businesses, we've learned the fundamentals of growth, and in a way we're already growing.

The type of growth we're talking about involves continually becoming a better version of ourselves, and continually becoming something more. We learn to love acquiring self-knowledge, embracing our quirks, mastering our emotions, and sublimating our desires and lesser urges no matter what condition we find ourselves in. As we saw in the health section of this book, the key to growth during uncertain times is to practice mindfulness with each thought as it arises. Either our thoughts are useful and will help us, or they are not and we must cast them aside. By picking and choosing the thoughts we focus on, we don't let our amygdala (the fear, animal urge, and survival center of the brain) dictate what we focus on. This type of mindfulness is a walking meditation we can practice day in and day out. Mindfulness is a practice which compounds and becomes easier over time. When we learn to watch and disassociate ourselves from our thoughts as they arise, we are ensuring that we empty out our mental trashcan as we go.

The next layer of growth is a daily focus on the HRSE practice. Can we do a small thing each day to positively impact our health, build our relationships, hone a skill, or become more entrepreneurial?

In order to truly master growth, we need an ultimate ideal or pursuit so we know why we're building, learning, and becoming amazing human beings.

The best definition and word for that ideal means revisiting the stoic pursuit of apatheia. It's the ideal mental state of continual mindfulness where we've valued everything and everyone in our lives as objectively as possible. This pursuit of apatheia is purposely difficult for our human minds, and that's the point. Pursing apatheia wrenches us out of comfort and complacency. It's often only through pursuing what initially feels impossible or incredibly ambitious that we are able to achieve anything meaningful. It's through spurring ourselves on and never accepting the illusion of defeat that we can get closer to the ideal state of apatheia where life is a thrilling journey. It doesn't hurt to always have a bit of delusion in our ambitions. Delusion in our ambitions is one of the most prudent teachers. When we're delusional, we set our ambitions too large, make the effort, and sometimes fall short. At this point, our previous state of delusion makes it abundantly clear what we need to learn or possess before attempting such a pursuit again. Delusion in isolation or life and death circumstances is dangerous. Delusion after we've mastered the art of survival is the foundation of greatness. Just the right amount of delusion and determination is the difference between a life that's a great story, and a life where we tell the stories of others.

Growth means we are always practicing building and creating value. If we haven't yet mastered the practice of receiving, asking for, and exchanging money with others for services or products we render, it is impossible to become entrepreneurial or ever start an impactful business. An online business doesn't mean a thing if it's the 100th Internet marketing blog started today. We don't grow when we do the same things as everyone else. Just because people tend to gravitate towards what's easy doesn't mean we have to. Growth is not easy, and it's not pleasant while we're in the trenches.

Fate can be the harbinger of excuses and quiet lives. Fate is what turns the optimist to cynic, or critic. Or, fate can bring us the chance to take massive action when the right opportunity presents itself. The right opportunities will be evident to those who have mastered survival and growth. Once we've mastered survival and growth, we can move onto what the Internet has made easier than ever: signaling.

Signal

The Internet offers us the potential to create and sell things (and opportunities) which don't require us to be physically present to recognize them. When we signal for what we need and want in a way that sells or provides something of value to others, we'll receive resources, jobs, money, opportunities, etc.

To learn how to signal, we need to understand three concepts:

1. Opportunity generation
2. Social proof
3. Exposure (How to close the sale and create what looks like luck from the outside.)

Opportunity Generation

This term can be summed up as how we present ourselves online. What platforms do we have a social presence on, and what does that presence communicate in terms of the opportunities we want? What do the text, bullet points, and our website say about what we're looking for? What type of value proposition do we offer to those we want to exchange value with? The idea is simple; we showcase our knowledge and expertise in a way which sells us to decision makers, customers, or whoever views our online presence.

Social proof

The next part of signaling and creating opportunities involves demonstrating social proof. How do we "prove" our abilities to others using commonly recognized social cues and credentialing bodies? Are we signaling relevance to the people or opportunities we want to attract? Are we using words and credentials, or citing verification of our capabilities by third parties whom the person viewing our social presence will recognize? Does the method we showcase our competencies give the person reading them a reason and an easy way to take the next step and contact us? Are we signaling and portraying social proof in a way that gives them a reason to respond to the emails we've sent them?

Exposure

Now we can become proactive in reaching out to others. We can begin thinking about how to accelerate opportunities through giving our online presence more exposure to those we want to send a signal to. We can gain exposure and attract interested parties through the tweets we're writing, LinkedIn articles we're publishing, status updates we are posting, the book reviews we are sharing, the emails we are sending out, the videos we are creating, the interviews we are giving others, and many other ways as well.

The Internet and technology have lowered the costs to signal, effectively making each of us our own media and public relations team. The Internet is continually moving us towards an environment of more perfect signaling. As we saw before with the mission of LinkedIn, they're trying to facilitate perfect economic signaling.

By beginning now, and setting up our presence online, we create an opportunity generator for ourselves. Then, we can perfect the social proof and give ourselves the exposure we need to attract what we want.

Strategy and Action

"If you're going to try, go all the way. Otherwise, don't even start." –Charles Bukowski

"If you take risks and face your fate with dignity, there is nothing you can do that makes you small; if you don't take risks, there is nothing you can do that makes you grand, nothing." –Nassim Taleb

Stay Alive, then Thrive

These are some extra ideas, actions, plans, and mindsets to help build our ability to survive, grow, and signal.

Survive

Survival is completely underrated. Think about how many people are dependent on a single source of income. They have 100% of their eggs in one basket, making them completely fragile to what the future and fate will bring them. The questions below are for us to answer, to better cover any gaps in our survival strategy. In order to bolster our ability to always thrive economically, we must start asking better questions about how we can diversify our income.

If we lost our current job, where could we earn other income? How could we develop an extra source of income today? What is the smallest step we could take today to build another stream of income?

How could we double our current income through multiple

sources? How could our skills be combined with those of someone we know to multiply our income? How could we partner with someone who already has a great product or service and sell it for them? How could we find others who can help us sell our products and services?

Could we develop another income source with our spouse or friend in a week? What if we *had* to? What about in a month? What if we *had to have another source of income tomorrow*, what would we do?

Debt in liquid credit markets is a tax on impatience. How can we immediately limit or pay off our debts? How can we become more patient and stop taxing ourselves?

How can we increase our savings? What purchases are we currently making that don't help us create more wealth, options, or enjoyment in our lives? Where can we prune spending?

Where do we spend our money to make more money? How can we purchase assets, instead of buying things that depreciate?

If our time is worth $50 an hour doing the work we currently do, can we hire someone to do the menial tasks around the house so we can work an hour or two longer every day? What are we doing now that can be outsourced tomorrow? How can we immediately raise our income by doubling down on areas we're skilled in?

Who has mastered the art of survival that we can emulate? Who has multiple income streams? Who is that person we know who is always getting job offers and opportunity? How can we learn from them? Think of who these people are and reach out to them or study what they do. If they're consistently getting great jobs, making money on the side, or doing interesting stuff like working remotely while traveling, how are they doing it? Find out how they are doing it, then learn and emulate them to gain extra income sources.

Try making a HRSE tracker and for 66 days record something small you do each day for each category. Behavioral psychologists tell us that if we can do something for, on average, 66 days, we have a high probability of forming a habit. That's where it becomes necessary to track our progress and make the HRSE practice a daily habit.

My favorite tracker for to-dos is the humble 4x6 note card. Grab a stack and take a card with you throughout the day. Below are some examples of things which can go under each category:

Health: Planks, Dead lifts, Squats, Sprints or whatever exercises give us the highest yield.
Relationships: How did we grow or make our relationships better today?
Skills: What did we work towards learning today?
Entrepreneurship: How could we add a few extra dollars of income to our bottom line today? What is one small step or bit of information we need to learn to get started? Can we hire that person who has achieved what we want via a service like Clarity.fm? How can we get started right now, to master survival? What is one extra income stream we could start building today? Don't worry if the extra income stream is really small at first. An extra $10 a day is $300 a month- enough to cover an entire repertoire of learning material to help us break into learning even more high value skills.

Just think about what an extra $10 a day could materialize:

- Smartphone: A high quality computer with Internet connection and a data plan: (approximately $100 /mo). This comes with podcasts where we can start to learn and find digital mentors.
- Kindle Unlimited: All the books we can read ($9.99/mo)
- Amazon Prime: Cloud storage, instant video and music, and delivery of anything we need in just two days ($8.25/mo)
- Audible: Audio versions of books on our Smartphone to learn while we walk, workout, do dishes, etc. ($14.95/mo)
- Food: Healthier grocery list including chicken, coconut oil, and vegetables ($181.75/mo)

This is an over simplification, but also a recognition that we can do a hell of a lot with an extra $300 a month, or $10 extra per day.

That extra $300 a month might come by listing the skills you have which people would pay for. Email somebody who needs that skill performed. Tell them they can pay you through PayPal and then deliver your skill. Voila! Instant income. To go more in depth and spark the idea muscle, feel free to use and implement any of the ideas from the ***75 Businesses to Start Today***. This is a list I created to help you get started at mastering survival, growth, and signaling right away. This list of 75 business ideas can be found at the end of next chapter.

The survival practice takes the most focus because it's the most difficult. Many people think they have this part of the equation mastered, but they only have one or two income sources. It's important to continually ensure our foundation of survival before we advance into growth and signaling.

Grow

After our survival is mastered, bills are paid, and income is coming in, we can begin to really dive into the growth part. Some questions to ask ourselves along the way of building a growth practice:

How are we growing now? What areas are we becoming better at? The best and most comprehensive test to see if we're growing is to look back at who we were six months ago. Are we embarrassed by things we thought, did, or were pursuing? By embarrassed, I mean are we aware now of how little we knew then? If we always look back at who we were in the past and find a weaker, less competent version of ourselves, we ensure that we're growing. It's important, of course, that we develop the ability to be proud of who we are today, yet laugh at ourselves six months in the future.

Signal

Signaling is about capitalizing on the powers of technology. The purpose of signaling is to showcase our work and attract the right opportunities. We want to use the Internet to help find and source what we need for our pursuits.

If we need a new job, a co-founder, or funding for a business, we want to get into the mindset of signaling for them.

For instance, if a job or better job is the focus, we want to signal via our LinkedIn or AngelList profile. Then we let those optimized profiles work for us, and present themselves to the thousands of recruiters scouring on those platforms. With a mindset of being focused on signaling, we begin to do the right work instead of becoming busy, endlessly sending out resumes to those who don't want them. Developing a focus to continually call ourselves out on how we're signaling means we do work that scales, instead of work which always requires we stay busy.

If we need a co-founder for a startup, how are we signaling? What Meetup groups are we a part of? Do we have a profile on CoFoundersLab setup? Is there a LinkedIn group where a co-founder might be found?

If our business needs funding, what are our sales? How are we showcasing those sales in a way that the investors we want can browse them at their convenience? If the answer is, "We need funding to build it," that's usually the wrong answer. It's an answer which will crush our ability to ever build a business. Most often it will attract people and investors who are a bit too much on the delusional side of the equation.

When we survive, grow, and then signal, we take full advantage of the audience and opportunity the Internet provides. Now, let's examine 58 virtual mentors who can help guide our actions along the way.

58 Virtual Mentors

"The mediocre teacher tells. The good teacher explains. The superior teacher demonstrates. The great teacher inspires."
–William Arthur Ward

"I not only use all the brains that I have, but all that I can borrow." – Woodrow Wilson

The Best Mentorship and Inspiration in the World

Mentorship is one of the most misunderstood concepts in the professional world. Sheryl Sandberg, the COO of Facebook, has a chapter about mentorship in her excellent book, *Lean In*. On mentorship, she says, "Searching for a mentor has become the professional equivalent of waiting for Prince Charming." She couldn't be more right. Many people feel entitled to a great mentor; they make the search for a mentor in person, instead of recognizing a few simple truths:

There are so many teachers in the world today and respecting and treating existing ones with dignity is fine. Treating everyone and every moment as a potential teacher is powerful. But we should also seek out the people who are the best in the world at their craft, and from whom we can receive virtual mentorship.

If someone is creating immense value in the world, it's rude to ask them to mentor us directly. Their time is the only commodity they don't have more of, so asking for them for time without compensation is incredibly selfish (and likely to fail).

We have access to the best virtual mentors in the world, and we're completely free to learn from them while they work. We can study their businesses, listen to their interviews, read what they write, and receive priceless lessons. This means we don't have to take away from their time creating while we learn from them.

"If we study what is merely average, we will remain merely average."
–Shawn Achor

So without further ado, here is a list of virtual mentors and curated information sources for those who want to rebuild America:

The Art of Business

Jeff Bezos: Follow his portfolio of businesses which many just know as "Amazon." These businesses are in varying stages of maturity and profitably. There are business lessons upon business lessons to learn from the stories behind the story of Amazon. Genius. Watch for same day delivery on everything. Look underneath the technological infrastructure running everything to see the handiwork of Amazon Web Services. Amazing. I'm glad he never got funding for one of the first businesses he worked to create: a subscription service which sent news to people via fax machine.

Elon Musk: SpaceX, Tesla, and Solar City. Lessons stacked on lessons in examining the history of his companies and the moves he's made. Musk is an incredible leader and salesman in addition to his other accolades.

Nassim Taleb: Read and re-read his *Incerto* book series and his social media posts via Twitter and Facebook. If you're up for the challenge and want to dive into the math behind his books, read his academic papers. http://www.fooledbyrandomness.com

Peter Thiel: Read, and re-read *Zero to One*. http://

www.foundersfund.com

Marc Cuban: http://blogmaverick.com

Anthony (Tony) Robbins: Pretty much everyone on this list has learned, studied, or indirectly been taught something by Tony. http://www.tonyrobbins.com

Gary Vaynerchuk: GaryVaynerchuk.com

Chris Sacca: I want to learn from anyone who was $2.1 million in debt at the age of 25, and 10 years later becomes one of Silicon Valley's biggest angel investors. Sacca fits the bill, and his funds now manage well over a $1 billion. http://lowercasecapital.com

Jay Abraham: One of the greatest marketing masterminds ever. http://abraham.com

Seth Godin: http://sethgodin.typepad.com

Shark Tank: Watch the show for an enjoyable kickback session which subtly teaches entrepreneurship.

Richard Branson: A master at branding, advertising, and crafting deals to mitigate risk. You probably already see him everywhere.

Ben Horowitz: http://www.bhorowitz.com

Travis Kalanick: Study Uber.com for the execution and creation of a global monopoly in the face of regulation. They may push too hard in certain situations, or it might just be well executed business genius.

Health, Relationships, the CEO Practice, and the *Good Life*

Dr. Rhonda Patrick: Personal health is the foundation of everything else, and Dr. Patrick articulately lays out the best prescriptions, diet, sleep, exercise, and a long term focused view on how to live a healthy life. She is a practitioner, a scientist, and generally a doer. FoundMyFitness.com

James Altucher: The James Altucher Show, Ask Altucher and http://JamesAltucher.com

Kamal Ravikant: http://www.evolve.vc, http://www.amazon.com/author/kamal

Tim Ferriss: Tim Ferriss Show, http://fourhourworkweek.com/blog/

Stefan Molyneux: Freedomain Radio, https://freedomainradio.com

Marc Maron: WTF Podcast http://www.wtfpod.com

Dave Asprey: Bullet Proof Executive, https://www.bulletproofexec.com

Economics, Politics, History, Culture, and Long Term Thinking

Thomas Sowell: Every book he's ever written is immensely valuable. This is what real economics looks and sounds like. It's immediately applicable to help create a better world and a better life. http://tsowell.com

Dan Carlin: Teaching Vehicles: Common Sense, Hardcore History, http://www.dancarlin.com

Brett Kelly: The Art of Manliness, http://www.artofmanliness.com

The Long Now Foundation: LongNow.org, their podcast: Seminars in Long Term Thinking

EconTalk: EconTalk.org and George Mason University have a fantastic podcast with guests covering a broad spectrum of business.

Virtual Mentors for Those Aspiring to Break into Politics

Justin Amash: http://amash.house.gov
Rand Paul: http://www.paul.senate.gov
Ron Paul: Peace and Prosperity, http://ronpaulinstitute.org
Rush Limbaugh: http://www.rushlimbaugh.com
Glenn Beck: http://www.glennbeck.com
Heritage Foundation: http://www.heritage.org
Cato Institute: http://www.cato.org

Get Your Hands Dirty, Making, Tinkering, and Building

Cool Tools: http://kk.org/cooltools/
Mike Rowe: www.proufoundlydisconnected.com
DIY Drones: http://diydrones.com

Technology

Steve Jobs: Read the Walter Isaacson biography Steve Jobs. There is also plenty of unique perspectives in his past interviews. Steve Jobs was what Peter Thiel would call a "definite optimist" in that he envisioned the future and didn't ask others if they agreed with where things were going. Instead, he just got busy making what he envisioned. One of the best case studies in the world for creating products people didn't know they needed. Studying Jobs offers tremendous examples of how to pre-sell, design, fall, rise, and create technology products which help people do more with less. http://en.wikipedia.org/wiki/Steve_Jobs

A16z Podcast: http://a16z.com
Marc Andreessen: http://blog.pmarca.com
Naval Ravikant: http://startupboy.com
Brad Feld: http://www.feld.com
Mark Suster: http://www.bothsidesofthetable.com
Kevin Kelly: KK.org
Jason Calacanis: Calacanis.com, This Week in Startups
Steve Blank: http://steveblank.com
Paul Graham: www.paulgraham.com/articles.html
The Signal vs. Noise blog of Basecamp: Incredibly valuable for any business, https://signalvnoise.com
Kevin Rose: http://foundation.bz
Noah Kagan: www.Okdork.com

Helpful and Useful News

Create your own stream of those you want as your digital mentors: Feedly.com
Wired.com / Inc.com / FastCompany.com
Quartz: http://qz.com
Sarah Lacy, Pando Daily, http://pando.com

Creativity and Art

Brain Pickings: http://www.brainpickings.org
Steven Pressfield: http://www.stevenpressfield.com
Srinivas Rao: Unmistakable Creative, http://unmistakablecreative.com
Hugh Howey: HughHowey.com
Evan Williams: Medium.com/@ev

19

From Entrepreneur to Founder

"Freedom lies in being bold." –Robert Frost

Above all, we must realize that no arsenal, or no weapon in the arsenals of the world, is so formidable as the will and moral courage of free men and women. It is a weapon our adversaries in today's world do not have. –Ronald Reagan

The 52nd Attempt

In 2001, three students won a competition at their university to build a video game. Then they sold that first game to a larger business for a modest profit. After that first small victory, no more would appear for several years. During that quiet period the team practiced their craft, acquired skills, and continued building games. The three team members kept working until 2005, when they were able to land their first round of angel investment. They kept building games and, by this time, were becoming quite good at it. They had built 51 games, and had started to publish some of them on the brand new Apple App Store, which looked like it could finally solve their distribution challenges. Another year passed. It was now 2009 and the team had been at it for years. They decided to keep building and the team pulled together to create one more game, which would be their 52nd. The Apple App Store was now starting to heat up and the game they released started to get thousands of downloads. Pretty soon Apple would feature their 52nd game called Angry Birds. After that, Angry Birds would go on to net billions of dollars and launch a massive business for its founders. The founders didn't just try a few things. They perfected their skills for years. They kept putting work in a place where it could be exposed to upside. Their first 51 attempts contained the learning and skills necessary to ensure that their 52nd found the right signal (the brand new Apple App Store).

When many people start a business, or seek to become more entrepreneurial, they forget that dozens of attempts are the standard. It's only those who grow during the dozens of attempts and small wins that eventually get to the monumental victories. It's those who are willing to reach out and email hundreds of people who get a meeting. It's those who reach out and introduce themselves to hundreds of CEOs who get jobs. It's the salesperson who reaches out to 50 more people than her quota in order to make things happen. The more we are willing to endure multiple attempts (and rejections), the fewer people we're competing with. Eventually, our skills and expertise will have built up so much that we can't help but stumble on that magic number of attempts, whether it's 52, 65, or 105. It is those who can endure who will find the best victories.

When we execute consistently over the years, and achieve small daily goals, we can't help but move closer to our dreams. Along the way, we never know where our efforts will be aided by serendipity, the rewards from acting on faith, or by the eighth wonder of the world- compounding. When we think large, we gain enough energy to act locally. Aim huge, act consistently, rinse and repeat.

The beauty of becoming more entrepreneurial or breaking into technology is that we only need to be right once. This is the exciting nature of business in an interconnected world, where there are no upward boundaries. Three billion more people are coming online and will soon each hold a Smartphone and be ready to participate in the world's economy. Millions of these people are learning English right now so they can participate. All these people can access what we create, they can use what the startup we join sells. The Internet allows us the ability to offer our service, product, or technology in a massive $17 trillion national GDP pool and a $78 trillion global economy. A single victory is all it takes. Envisioning that victory can spark the doxastic commitment necessary to weather however many attempts it takes.

Strategy and Action

"Ideas are the beginning points of all fortunes."
–Napoleon Hill

"...the proper response to a lousy idea is not to stop thinking. It is to come up with a better idea." –Kevin Kelly

Ideation, Next Steps, and Giveaways

Becoming a founder with an end story the media champions doesn't come from just having a great idea, or even having hundreds of great ideas. Successful founders are confident in their ability to generate, execute on, or sell a massive amount of consistently great ideas. The only way to amass a large number of great ideas is through building up the practice of generating a lot of bad ones which we can't do anything with. The process might feel like day dreaming; it might feel silly in practice, but the true creative and lateral thinkers we read about in the headlines have become masters at ideation. It's impossible to become successful with only a few great ideas. If we're going to embark on joining a startup or starting our own business, we need to consistently generate great ideas, filter out bad ones, execute the ones we can, sell the ideas we need to other people, and give away the right ideas strategically. Just like a good product is dead on arrival without distribution, we must know how to train ourselves on to generate, distribute, kill, and build ideas appropriately.

This practice starts by listing ideas whenever they come to mind. It's important we write them down, whether it is on note cards,

Evernote, or whatever medium we prefer. We can list them as they arrive, or we can just get in the habit of listing 5 to 15 ideas every day. We will never get good ideas before we begin this practice. The time when we're most likely to be creative and get the best ideas is in the midst of writing down bad ideas. We can start ideating by listing ideas to help, fix, or improve situations immediately around us. Many of these ideas won't be good, and that's fine. But list them, rank them by order of how good you think they are, and then explore how to execute them. As you get good at implementing small ideas in your own life, branch out and help others. Think of new ideas for products, services, businesses, distribution channels, joint ventures, ideas for other people, articles to write, books to write, movies to make, business jokes, micro businesses, games. It doesn't matter what they are, the point is just to practice.

Ideate without keeping track. Just do it every day and don't worry about what ideas you write down at first. The best ideas will continually surface and then resurface.

When we start finding ourselves coming up with really good ideas, we can then go about determining the next step to take to make our idea a reality. If we can't figure out the first step, there might be somebody else who can. Maybe we have an idea which would be perfect for someone else to do, in which case we can reach out and give the idea to that person. This is the start of many business relationships and friendships, and gets the karma ball rolling.

If an idea seems really amazing, get started on it right away. Consider what is required to take the first step today. Ideally, pick the hardest step, like asking somebody to buy a product or service before it's ready. As long as we're sure we can create what we sell, there is no shame in pre-selling great ideas. It's important to avoid doing useless things at this stage. (An example of a useless thing: filing for an LLC before we have paying customers.)

We don't have to worry about formally starting a business or finding a co-founder. The best ideas are ones we can start dabbling with, with friends. From there, we can let the idea breathe and see what type of business potential it has. The hallmark of great ideas are ones which quickly improve the quality of lives of the people around us, or in the case of business, are something others are ready to pay us money for. If our friends or family purchase what we're selling because they genuinely want it, we could be on to something huge.

Co-Founders and History > Mystery

Most ideas we develop will take a team or small group of people to test. If every part of our idea can affordably be outsourced to freelancers, we can try this, but for larger businesses we'll eventually need co-founders or employees. Start looking for them among existing friends. Don't think that the best business partners are waiting at networking events or are recent graduates from an MBA program. As beginners, we shouldn't expect people we've just met to become amazing business partners. If we were developing ideas which could eventually become a big project or business, ideally we'd only want to work with people that we have a history with.

You Can Start Alone, but You Need a Team to Finish

Find a co-founder if need be. Break down the idea with them and get started. Ideally, the work you do by yourself and signal via your online presence will work for you to signal the right co-founder, at the right time.

Sell the By-Products

When you're ideating, many of the ideas will be great for people with established businesses. We should feel free to contact people who could best use our ideas. This is a great way to practice improving our ability to actually come up with useful ideas. Any ideas we have which we can't use, we should give to others. Give often and give freely. The less work the ideas require to execute for the people we deliver them to, the more generous we're being. Creating more emails for someone isn't useful, but giving them a carefully researched, "I saw you did (this) and I thought if you did (that) it would help you get more of (what that person wants)," would certainly be helpful Become known for being creative and helpful. We can try different approaches to see what types of email, proofs of concept, or offers it will take to get people interested in working with us. We can reach out to veterans who live close to us via LinkedIn, Twitter, or Facebook and ask them to work with us.

The practice of building ideas has been popularized many times. Earl Nightingale started championing the process decades ago, and James Altucher recently breathed fresh life into the practice. The practice of generating ideas is tried and true, and can quickly lead to that extra $300, $1,000, or $100,000 a month in income. So in the spirit of generating and executing on great ideas, let's now move on to: *75 Business Ideas* we can start today.

75 Businesses to Start Today

"A journey of a thousand miles begins with a single step." –Lao Tau

"The whole thing seemed pretty iffy at that stage, there wasn't really anything except for a guy with a barking laugh building desks out of doors in a converted garage"
–Shel Kaphan (first Amazon CTO- on joining Amazon with Jeff Bezos)

Caveats & What Early Business Starts Look Like

The term "business" can be tricky. Businesses don't start when you get business cards, or file for an LLC, or plan over coffee. Businesses officially begin when we receive money from someone in exchange for a good or service.

Thanks to the Internet, there are more ways than ever to learn about entrepreneurship. But the real learning begins when we set up stakes and incentives, and squeeze ourselves until we take action.

The list below is of 75 businesses we can start today. The goal of these business ideas are to help move us as close as possible to making our first dollar. All these ideas are just starting points. They can be modified (and should be) however we see fit. The goal is to just begin, and get our entrepreneurial muscles working immediately. We can look at the ideas below as businesses, experiments, micro-businesses, or incentivized experiments with stakes. What matters most is that we get started, so we can get learning.

A Caveat about the Sharing Economy

The mainstream media and most of the technology industry love the phrase "sharing economy." This is annoying on a few different levels, mainly because we're not sharing anything; the market for everything has just become more liquid. Technology has lowered search costs and allowed people to do business with one another much more easily. These people aren't sharing anything, they're buying and selling goods and services (albeit temporarily). Just because the mainstream media turns its nose up to what business looks like on a small or local level doesn't mean we have to use the language and terms they give us. The "sharing economy" is a blessing to anyone who wants to build their entrepreneurial muscle. It's an opportunity for each of us to start businesses more easily, and create value in the economy for everyone else along the way.

The platforms and business which have helped make the economy more liquid such as –AirBnb, Lyft, Uber, or even DogVacay, allow us to instantly become more entrepreneurial. We can begin the learning curve today and get paid to learn the right things. Opportunities like these to start businesses more easily lie in stark contrast to most MBA and entrepreneurship programs where we pay to study often outdated business case studies. This formalized style of education and college training is certainly socially acceptable, and perhaps easier to explain or justify to our family. But all the ideas below put us on the path of real learning. They all are designed to spark ideas or invite us to take the real actions necessary to learn about business and entrepreneurship. The best lessons and learning we'll always remember comes only through practicing. All the 75 ideas below are about practicing and about making money, instead of paying to study old examples.

Many people get caught up in the humble nature of many of the business ideas below. This is natural and to be expected. After all, traditional media focuses most on the end polished products of entrepreneurs. They often leave out the humble beginnings that all of us are capable of beginning today. To help prevent the humbleness of the ideas below from stopping us from taking action, let's take a moment to reflect on how noted entrepreneurs got their start in business.

How Did They Get Started?

"That's sort of the beauty of creating it. It often takes the act of creating it for the idea to emerge. Often it's not there at the very beginning." –Kevin Kelly

Aaron Levie, CEO of Box.com: Started many small web design businesses.

Tracy DiNunzio, CEO of Tradesy.com: Rented out her apartment via AirBnb to make extra cash while she was learning and building what became Tradesy.

Ilan Zechory, Co-Founder of Genius.com: Started selling his services as a licensed hypnotist before co-founding what became Genius.

Richard Branson, Founder of The Virgin Group: Started selling advertising space in a magazine he started before the magazine actually existed.

Marc Cuban, Founder of many companies including Broadcast.com. Now Cuban is a VC on Shark Tank, Creator of CyberDust, and Expire, among other products and apps. Cuban got his start in business through bartending, selling powdered milk door to door, and then learning about software by selling software. After he got fired from his job selling software, he started a computer servicing company to fix all the problems he saw at his previous employer.

Tim Ferriss, Author, and Angel Investor: Got his start with a conference for speed reading, and an audio program about how to get into an Ivy League College; then he started a business selling a single workout supplement through website, phone, and mail orders.

Travis Kalanick, CEO of Uber: Got his start with a business he founded with a neighbor. They tutored large groups of students for the SAT.

The examples of small starts are everywhere. The difference in those which start small and stay small versus those which start small and go large is only mindset.

When someone pays us for a product or service we render, we make his or her life better. At that moment, we are doing our part to rebuild America. That gap between the costs of what we create and the value it brings to someone else (our margin) is real philanthropy at its core. There isn't anything more American than creating value where none previously existed.

Creating businesses is also a fantastic mental and financial insurance policy. When we know we can start with nothing and build multiple income streams, we gain a rare type of confidence. There are few things more satisfying than knowing we can lose everything we have, and find a way to make money.

We open more options to ourselves by cultivating the ability to start businesses. Once we've built that entrepreneurial skill, we can be dropped anywhere in the United States, and within a few months be up and running, making money by moving resources from areas of lower yield to higher yield. This is the ultimate optionality and freedom. There isn't a better path to crushing our own fears.

There is no reason to wait. The ultimate foundation of business skills starts with many small ideas like the ones listed below. The real learning lies in doing things that require hard work, courage, and bravery. An example would be looking someone in the eye and asking them to write you a check, or send payment for $495 for your services.

75 Businesses Ideas We Can Test Today

Sell your unused clothing (or your spouse's) on Tradesy.com. Set up a consignment shop and sell clothes or other goods online. Practice selling popular apparel on eBay or Amazon. Focus on making a small margin at first, and gradually increase it.

Start your own service hiring, staffing, or lead generation business. Find businesses which need specific jobs filled. Then go on LinkedIn and introduce qualified candidates for them. Staffing is a multi-billion dollar industry, and extremely lucrative in many fields like technology. In addition to staffing, you could help bring in leads for attorneys, car dealerships, or for higher education providers.

Write non-fiction guides on Amazon to answer the questions people are searching for. Study commonly asked questions on Quora to get product ideas. Study the "most helpful" three star reviews of existing non-fiction books. Find ideas to create better non-fiction books to answer the most important questions people ask. You could put all these books under a brand, or even a pen name if you're worried about how your writing will be perceived. You could even start by offering summaries of business books, how-to stuff, or offer your personal answers to problems you've solved in your own life.

Thanks to KDP and Kindle Unlimited, it's never been easier to publish and get distribution.

Write fiction on Amazon if you're creative. Either go into KDP and Kindle Unlimited or upload to Smashwords as well. Serialized fiction sells well. If you love to write there has never been a better time to do so. Remember, though, selling books is still selling. Without a publisher behind you, you have to do the marketing. There are articles and eBooks to tell you how to do this, but it does require commitment and hard work.

A Writing (Content) Company: Start producing content for blogs, then outsource the writing and just edit it. Top __ lists are dead simple. Read James Altucher's *30 tips for Better Writing* and you'll be off to the races. Find a few people or content sites that can pay you for what you write. Start off writing for those with a high ability to pay, who can benefit from making a single sale. Create a profile or website for yourself and services. You can eventually build this into an agency. Maybe a personal brand will work initially to find clients, or maybe the appearance of an agency allows you to charge more. After you get started with a few paying customers, you can create profiles for your services on many existing sites and marketplaces. All of these profiles will be used as lead generators to bring you clients. Consider: Contently, Clarity, Thumbtack, and even Fiverr. All these profiles can be turned into well oiled machines bringing in a stream of clients. Expand into offering other forms of content creation as your margins allow. This small start has potential to be turned into an agency or marketplace for content creators.

Graphical Content Creation Company: Master Canva, using it to create graphics, infographics, Twitter cards, Facebook, and Pinterest content for businesses. Grow the business or a find a niche opportunity in an industry that you love.

Flipping (insert decently priced electronic or mechanical tool here). Take cars as an example. Can you clean up and fix cars? If you can do the work necessary to clean or fix those older or broken cars, you can buy and sell them. The lifespan of cars (and many other electronics or mechanical tool out there) is rapidly increasing. The influx of 68 million immigrants to the U.S. between now and 2050 means there are millions of people who need low-cost transportation, electronics, and tools. This is a great near term start to begin making cash if you're mechanically inclined. If your prospective buyers don't have enough cash to pay you, simply negotiate a deal with a local financing company or bank. You refer the customer who wants to buy what you're offering to them. If the margin is slim on the products you're flipping, you can still make money for sending the financing company the customer lead. In fact, consider the fact that new banking customers are worth thousands to tens of thousands of dollars to banks.

Events, Conferences, and Meetups:

Traditional retail is dying. E-commerce and expedited shipping is exploding in popularity and demand. The retail stores which survive over the long term will gradually turn into event-based/destination-based places where activities happen. If you look at many Whole Foods locations, they teach cooking and craft classes at their stores. If you look at Lulumon Athletica (premium athletic and yoga gear), you will see that many of their stores have yoga areas where they hold classes. Bonobos apparel stores exist for customers to try on clothes, be sized, and catered to. The customer orders the clothes from that point. The point is, every retail business needs to transform into an events and activities business. There is a business to be started helping other retail businesses get started in this transformation.

More Event Based Business Ideas:

Veterans Event: Host an event, conference, or meet up for veterans. Charge a small ticket price to keep freeloaders out. Offer connections to employers. Call employers who have veteran-hiring initiatives and see if they'll sponsor your event. Do a joint venture and get started creating or hosting events with a company like HirePurpose. Companies like HirePurpose already have the existing relationships set up with the content to attract customers, and the employers who need talent. Act as the intermediary and help them create a great event to bring all parties together. There are many other businesses like this in other verticals where you can create similar events.

Wine tasting at local restaurants, bars, or event venues. Find locations which have downtime or times that are not as busy. Offer to bring them customers in exchange for hosting your tasting. Contact 10 vineyards and get bottles of their wines to use as tasting samples. Offer a special deal to those who attend and want to purchase wine. Host the tasting! You can sell tickets, you could sell wine from the vineyards, the restaurant you host it at could pay you for bringing in customers, or you could start hosting tasting tours to several local vineyards (who could also pay you).

Craft beer making and tasting. Put up some flyers that say only, "FREE BEER" detailing the tasting event time and place. Or, if you want to teach groups how to make beer, put up a sign that says, "LEARN TO MAKE BEER", and "GET FREE BEER". Find a local brewery, bar, or venue you can partner with. If people don't want to pay for tickets, ask the venue or brewery if they'll pay you just for bringing people in. If you can bring 100 people to a bar or venue on an off day or time, I strongly suspect they'll give you a nice cut of the proceeds.

Craft Brew Pub: Start or join a local spot that has sprung up and get in on the ground floor. Start hosting events for them and become an expert at bringing in business.

Relationship Building Events: Introduce guys to women (or vice versa) and help them increase their relationship skills. If you're a woman, host an event for guys to get practice with talking to women. If you're a guy, help women meet guys who aren't assholes. Introduce women to guys and help them see that there are nice guys around. Author and entrepreneur Michael Ellsberg was able to host (and monetize) "eye-gazing parties" where people met up and just stared at each other. There is no reason you can't host a meetup event for those looking for relationships. Offer people an opportunity to practice improving their talking and relationship skills in a place where everyone around them is doing the same. This type of positive social pressure can help many of them overcome their fears.

Forced Public Speaking Event: Most people are terrified of speaking in public, but many want to learn. Host an event which helps people learn to speak in public by speaking in front of the attendees of the event. Don't allow phones or pictures inside the event, in order to give people a comfortable place to practice. This business idea could be extremely lucrative if executed well. You can create an event where you have a day of short, structured talks where people get on stage and speak many times in front of conference attendees. You could start off by having them walk on stage, introduce themselves, then walk off with everyone clapping. Have everyone cycle through, and gradually increase the time each participant speaks until they're speaking for 5 or 10 minutes each on a specific subject they know about. You only need 50 people paying $97 each to clear $5000 in revenue. In order to get better at public speaking, people need to have incentives and stakes, and this means building up series of successful small "wins" in talking in front of a large audience. Your event can be the place where they learn.

Local Industry Expert Event: Gather 20 local people who are industry experts in a very specific niche. Maybe they have a book, or maybe they each own similar businesses in the same vertical. The idea is to find people who own businesses which could do more business if they joined forces. Create an event to help introduce them.

Create a mastermind event which helps people learn to do or accomplish something very specific. You can find several people, or ideally it will be you yourself who has achieved what others want to pay to learn. Sell tickets to those who want to learn what each of the experts at your event has done. If you don't have a way to get the word out about your event, find somebody who does and split the ticket sales with them. For instance, many veterans want to break into law enforcement. You could very easily get 20 federal or local law enforcement experts to prepare a quick 20-minute presentation about "How to get where I am." The law enforcement agencies might even be able to sponsor or pay you to host the event. This is just one example, and there are opportunities to do this in fitness, diet and health, real estate, technology, and any discipline people want to learn.

Break Into Technology Event. A huge and growing number of people want to learn how to become developers, programmers, designers, and generally just become more competent with technology. Many people want to become involved with startups but don't know how. Log onto AngelList or search the INC 500 and find companies nearby needing to hire people. Find out what and who they need, and what they're willing to pay to find or train qualified candidates. You could start by hosting an event for all the local consulting companies who want to hire IT consultants. Find 10 to 20 local consulting companies and ask if one person from each of their companies would create a short presentation for those interested in breaking into IT consulting. You have your event content, now just sell or offer tickets to local groups. You might start at a local community college or university. If you find that many of the people signing up for your event don't have the necessary credentials, offer them training. Reach out to local training providers like developer boot camps, or IT training companies. Ask them to attend as well.

Host an event for two companies or individuals which should have a partnership. An example would be to promote an up-and-coming company which is disrupting an industry. The up-and-coming company is likely completely changing an industry, or cutting out the middlemen. An example might be Redfin and frustrated realtors. Redfin (or whichever competitor wins that space) will take away much of the market share from existing realtors. There are many frustrated realtors right now who need more business. By helping them understand how Redfin is changing their market, you're serving them. They might want to work for Redfin, or they'll get ideas for how their industry is being transformed, or how they can become more skilled. After the event, you'll probably stir up some opportunities and probably a few job or consulting offers (whether it's with Redfin or the realtors you're working with).

Another example might be hosting an event for college students or frustrated cab drivers who are interested in working nights and weekends at a great part-time to full-time job. Introduce them to driving for Lyft or Uber. Refer potential drivers to Lyft and Uber. Chances are you'll stir up an opportunity to work for one of those rapidly growing companies.

Host a hobbyist based event. Use space at WeWork, General Assembly, or a local incubator or accelerator. Start by asking 50 of your friends what they would pay money to learn or do. You could purchase a few cheap drones/UAVs and the event could be about how to use and fly drones. The idea is to see if you can create an experience those around you are willing to pay for. If you can create something good enough that those around you will buy (not just to be nice), you're well on your way to building the entrepreneurial muscle. The local co-working space will love the fact you're bringing in new people.

Grab, rent, or buy a small space of your own and make it multi-purpose: bars, breakfast, lunch, dinner, craft brewery, etc. Work at one of these places first to learn the business, or find what they need and provide it. Maybe they just need more craft beers. In that case, start a beer brand.

Consulting, and Agency Oriented Business Ideas

Help out at your local incubator or accelerator. The founder of that incubator or accelerator is probably incredibly busy. Start by helping out there. After you've delivered some value, find out what they need. Do they need new vetted companies to apply for the accelerator? Do they need new established companies to rent space from them? Find out their needs and deliver. Most companies working out of incubators, accelerators, or co-working spaces pay around $300 per person, per month. If they need additional companies to rent space, ask what they are looking for, what they'll pay you for, and what percentage of a contract they'll pay you in a fee for bringing the business to them. This is a great opportunity to be in a place where you can apply to their accelerator, join one of their portfolio companies, or get a job offer from the owner of the incubator or accelerator.

Consult with individuals and organizations to shrink their digital footprint online. This generally will decrease the amount of liability exposure they have. Small businesses will pay a consultant plenty of money to spend a day with them teaching them why they should use tools like DuckDuckGo, LifeLock, CyberDust, or Expire. This field will continue exploding in the future. For now, you can sell a service where you do the work of cleaning up people's online footprint or something similar. Create a course for individuals on how to protect their online identity or teach people how to shrink their digital footprint.

Work with existing Veteran Owned Businesses. Simply go to http://www.veteranownedbusiness.com and determine how you can work with or help these businesses. If anyone actually wants to hire veterans, it's probably the founders of these businesses. Talk to several of these businesses and find out what they need. Maybe they need more salespeople or a customer-service person. Find out what they need and bring that or be that for them.

Deliver value to VC-backed technology companies. Find top-tier VC backed companies. You can start by looking at the list of companies which are re-building America in chapter 15. Study their business models until you discover something they need that you could provide. Offer that solution to them, and see how they respond. Pester their CEOs or executives to figure out what they'll pay for. It may be more customers, an event, more hires, etc. Do your homework and research how you can deliver what they need. Deliver first, and then ask to get paid. This is how you can get your foot in the door and land the type of job or consulting gig which never gets advertised. Or, you'll build the goodwill necessary to be asked to apply if they do post jobs publicly.

Create video content for existing business. Many businesses are up and running, but need someone to turn what they do into video content. One example of a veteran-owned business which could use more great videos of their gear and events is GoRuck. GoRuck is in the events, clothing, and experience business. Study and participate in their events. Find out what type of pictures, video, or mini-documentaries you could create around their events. What if you created this content for them, and then sent it to the people who just finished attending a GoRuck event? This would be a great way to surprise and delight customers after an event ended. They'd open their email box to see a picture or video of themselves and their friends, showcasing their participation at the event. This is the type of content which would likely get shared. You'd help GoRuck virally market itself, and likely end up with a great gig or job in the process.

Start with simple video footage of the event, and detail each participant. The idea would be to create one loop and short promo film of everyone who participated. Edit it into short, thematic clips which could be easily shared with friends and family. There are many ways to expand this concept and help any event explode in popularity.

Create highly personalized and individualized testimonial videos. Get footage of satisfied customers. Introduce yourself and ask them some questions about their experience at whatever business you're creating the testimonial for. Record their answer, and add a branded logo of the company, product, or service at the end. If done right, video testimonials are some of the most underutilized marketing tools available to businesses. Instead of just more customers, testimonials help businesses attract the right kind of customers.

To piggyback our previous example above, at the end of a completed GoRuck event, each person who has just had a great time is probably ready to give a great testimonial. They've faced a serious challenge and will like to share this with others. This is the perfect time to grab a video testimonial from happy attendees. You, or the

GoRuck team, can deliver the short media testimonial afterwards. The participants will likely want to share it with others. Instead of charging for videos of their time at the event, now those who most enjoyed the events can help market the events. This is a type of growth engine or flywheel where everyone wins: customers get free videos, the business gets testimonials, and the videographer (you) gets paid.

This idea doesn't just apply to GoRuck, but any company in the experience business, which will be every retail and commerce company wanting to compete and stay relevant.

So find a company you like, help them create amazing digital content of pictures and videos showcasing what they do, and help their events market themselves. This could lead to an individual consulting gig, a job offer, or even the start of your own agency for creative services.

Drone / UAV Agency: The global demand for drones, drone pilots, and drone services will grow exponentially in the coming years. Being one of the first experts at flying drones and recording video for clients will be a huge business. Keep in mind that there are thousands of other needs and applications for drones other than video. There will be businesses which spring up around UAVs including new robotic attachments, modified sensors, custom built UAVs, and software and hardware to power them. All these opportunities are ripe to be developed and turned into a business.

Piloting and shooting high quality video footage from UAVs can open the door to a slew of business opportunities. Here are just a few businesses which will need UAV video services: cruise lines, residential realtors, bars, clubs, directors of films, commercial real estate, government, townships, farmers, hotels, retreats, national parks, marinas, and ski resorts. Every travel and vacation destination, every event held outside—everything from craft fairs to Tough Mudder style events will benefit from UAV films. Start the local UAV filming agency and branch out into every other application. The skies of the future will be thick with UAVs.

UAV Security Agency: Large facilities need security surveillance at reasonable prices. Chances are most of them already have security providers they're happy with, but you might be able to seriously undercut many competitors that have to use physical staff. Or, the UAV monitoring you offer can supplement the existing services they use. A single contract from a large corporation could launch a successful UAV security agency.

Consult on a site like http://www.Clarity.fm or LiveNinja.com if you're an expert in something.

Take the biggest pain point for everyone in your existing industry and figure out a solution people will pay for. If you need to build a product to solve it, team up with someone who can help.

Teaching and Training

Personal training for large groups of people all around your state. Instead of working with just single clients, teach only groups, only in big towns around your state. If you're close to a metro area you'll be able to easily sell and fill classes for 50 to 100 people at a time. You could do events which are single or even multi-day. For instance, you could sell a weekend training event where people get three workouts over three days, after which they'll need to recover for several weeks. Get creative and blend classes together that nobody else offers. For instance, offer kettle-bell classes combined with yoga and pilates. Or offer strength training with a blend of sprints and shuttle runs you setup. Offer partner dance classes which include a yoga portion. Create fun classes and teach them en masse. The personalized personal trainer only gets to work with one person at a time. Whereas these types of events will allow you the latitude to work with hundreds of people at once.

You could teach a course over a three-day weekend. Selling a single ticket which covers Friday, Saturday, and Sunday for two hours each day is an interesting way to structure your training. As the courses get good enough, you can even film and sell these sessions as you go, or just broadcast them for free on YouTube where you can advertise how to sign up or buy tickets. As you go along, splice together different sections of your work into a highlight reel or master class. Now you have a video product to sell on Amazon Prime Video or Vimeo.

Meditation, Mindfulness, or Technology Detox Retreat. People pay big bucks to learn about meditation and mindfulness. If this is something you're an expert on, consider teaching a single day, weekend, or even week long class or retreat to cover the basics. These retreats work because trying to get started on your own is notoriously difficult. Many people want to become better at mindfulness, as well as taking a digital Sabbath from technology. You can bill your retreat as a tech detox, as well as learning mindfulness and meditation. Some meditation-style retreats currently have "silent" portions where talking isn't allowed. These retreats are incredibly popular. Am I saying become a shaman and have people wonder if you've lost your mind? Maybe. How shamanistic you take your teaching is up to you. If you think old school religion is going to be radically changed, I'd agree. The Internet is getting rid of all the gatekeepers for many reasons. The priest or person who says, "I am the gatekeeper," is marginalized by the person who says, "I'm a teacher to help you see there is no gatekeeper between you and God/Infinite Intelligence/

Eternity/Surrender/Peace/Bliss."

Teach an online class about something you've done which others want to do. Sell the class on Udemy or Skillshare. Or sell your course through a tool like Fedora.

Become a public speaker on a topic people want to learn about. Topics people will always pay for better tips and advice on include relationships, making money, getting in better shape, and breaking into new industries. Find out what other people want that you know about and get up and talk about it. There is a huge demand from conference and event groups for public speakers who can actively engage with audiences. If you can think on your feet, make jokes, and teach while being semi-entertaining, there is great money to be made in public speaking.

Become a personal tutor for standardized tests, licenses, and certificates. If you can get a great score on a standardized test like the SAT, ACT, GRE, ASVAB, LSAT, or MCAT, and can break down how you got that score into a teachable format, there are thousands, if not tens of thousands, of people who would pay for that knowledge. Put that knowledge into a digital class, a book, a local event, or local test prep and help other people get great results on one of these tests—and get paid for doing so.

Become an online tutor. Record what you teach and offer the recordings for free online. Occasionally, package a bunch of your recorded lessons together and sell them as a course. A great example of this in a niche market is Sean Wes. He teaches people how to draw lettering by hand. If you're a creative type, this is a fantastic way to share and sell the by-products of your work in a way that gets your name out there.

Social Media Training. Help businesses learn how to use the latest social media platform more effectively. Ideally start by helping a business which is willing to pay or is willing to refer you to businesses that will. Start an agency that helps a specific type of business or professional on social media.

Creative Food for Thought

Food Trucks. These are springing up around Metro Areas and are much less risky to start than traditional businesses. Find a local Food Truck group that is crushing it and figure out what they're doing right, or even go work for them for a while. Either launch your own Food Truck from their brand or start your own Food Truck.

Niche down, and become an expert to the experts. Be the ___ Expert in an industry which is brand new and undervalued. For instance, learn everything possible about cryptocurrency. Write about it. Talk about it. There are dozens of venture backed cryptocurrency startups and businesses that need consultants or want to hire passionate people but can't find those competent with the issues.

Freelancing: Use oDesk, Elance, Thumbtack, Craigslist, or LinkedIn to start freelancing in whatever field you have marketable skills. You will become more of an expert and be able to charge more as you gain experience and accrue positive feedback.

E-commerce Stores and Opportunities

Create your own online school with Fedora. This web-app and hosting service for online schools allows you a complete turnkey solution to teaching classes. Get teachers, teach classes, or offer training that businesses want their incoming employees to have. Teach people skills, get paid up front, and then refer them to

employers and get paid for placing a candidate.
https://usefedora.com

Open an E-commerce Store. Want to sell t-shirts, posters or swag? Create a store with merchandise that people want to wear, simply by slapping a design, logo, saying, or catchy cartoon on a shirt. Start a store using a turnkey commerce solution like Shopify. Link Shopify up to fulfillment services which take care of all the printing costs, shipping, and fulfillment for you in exchange for a cut of sales. You can set all this up and have a beautifully-designed store via Shopify linked with Printful (which makes, prints, and fulfills orders for physical goods). You can setup a Shopify site in a few minutes, with a couple clicks.

Setup an e-commerce store to sell other people's products. Thousands of products are available in online databases. You sell them through your online store, and fulfillment is handled by another company. The margins are small, but massive businesses like Amazon offer this opportunity and get a huge percentage of total global Internet traffic.

Using Content Curation and Filtering to Deliver Value

Start a media buying agency. This starts by just learning Facebook advertising, then branching out to other platforms. There are millions of people in the world who need media and advertisement purchased who don't understand the latest platforms. The older platforms with older advertising products will offer little to no margin for users. This is where media buying gets lucrative because new or smaller platforms usually offer some arbitrage where the person buying the ad/media can win. This type of media purchasing is the new public relationships. When you buy ads or media, you either get a result or you don't. The old days of hiring a PR firm and never getting any metrics or deliverables are over. If you can consult with a business, then buy media to drive traffic, and add customers to their bottom line, your services will always be in demand.

Many companies have apps or services but have no idea how to reach customers on mobile. Teach them how, and acquire traffic for them. Once you learn enough about how to do so, your expertise can even open up opportunities to work for the platforms that offer the ad buying services you've been using. Or you could create a better service yourself to fix problems you encounter. Or, you could become or partner with a technical founder to create an advertising network.

Start a news site using a Wordpress news and magazine template, or Squarespace. Don't start a personal blog unless you're a creator, public speaker or CEO who will use that site to get clients. Starting a news site in an industry you want to learn more about is a turnkey way to open up the opportunities you want. An example might be in a growing industry like big data. You could start a news site on big data jobs, or big data news where you profile how to break into big data, follow industry trends, and link interested people with the training necessary to get started working with big data. Find existing bloggers and experts on this topic who already have content, and ask them if they want to repurpose it, or use it for your site. Create a job listing page from day one. Let the first 10 companies post their jobs for free. Go try to find them employees. Then charge the next companies that want to post jobs.

Other examples of niche news sites to start in growing industries: Drones/UAV, Bitcoin, Energy, Fracking, IT Security, Programming/ Coding, Consulting, Freelancing, Elder Care, Franchises. Find service companies which are leading the way in these industries and figure out the type of customers they need. Determine what types of questions customers who might use these businesses are asking. Start answering the questions customers have about the niche (or vertical) you're in, and point them in the direction of the service provider you partner with. This is a content site that can make money very quickly. You can monetize through affiliates, hiring, or selling digital courses.

Build a website geared towards teaching in a hyper specific vertical. Aggregate news and the latest details. Have pop-ups to grab emails. Create a great course for that space and teach and sell it through a series of email auto-responders. Or monetize via hiring, affiliates, etc. Bonus points if you can get writers or user-generated content.

Freelance

Freelance. You can use oDesk and Elance to get started, but the Eastern Europeans and many foreigners are already CRUSHING it on these sites. Or you can use Micro lancer or many others. Set up a personal website for the type of freelancing you do, and then put ads on Craigslist or local bulletin boards for your services.

Fiverr- Freelance services in bits and pieces with built-in up sells. Find out what people want, and offer that service. Have many levels of prices and offer different or higher priced services off the platform.

Researcher for an author: If you're incredibly detail-oriented and technologically savvy, reach out to an author who is publishing often and needs help researching topics. Demonstrate the level at which you can work and deliver. Work a bit for free, then charge. Build a portfolio of work and do work for several authors at once. Sell other services as the need arises; this is a bit like becoming the librarian of the future.

Selling: The Art of Shifting Assets from Low to High Yield

Setup an Etsy Store to sell your crafts or artisan creations. If you don't make crafts, you can set up stores for those in your area who do. Or find amazing products from the best artisans in a spatial area. An example might be Californian Artisanal Furniture or Amish Furniture (buy in bulk from the Amish and sell to Yuppies at salty prices).

Sell every single extra "thing" in your life that you don't use on Ebay and Craigslist today. Why? The quicker you become better at writing catchy ads and selling, the better you'll be as an entrepreneur. Minimize the stuff in your life. One of the ways to become entrepreneurial is to sell something— anything— as quickly as possible. The second is through decreasing the clutter or extraneous stuff in our life whenever possible. You can easily expand this business to home, attic, and garage, "minimializing" where you go in and sell all the extra things people have they no longer want.

Find something you love and sell that on a commission basis. There isn't a skill as valuable as making others' lives better through sales. This is the little-known most important secret in sales. The best salesmen sell only what they love and purchase themselves. Then all they have to do is to approach a company or person they buy from and ask to sell for them. No founder on earth would turn down a person who asks to sell a product for them because they love the product.

Liquidate Underutilized Assets on Service Platforms. It's not a sharing economy; it's an economy becoming more liquid thanks to people buying and selling with each other on these platforms! Find and liquidate underutilized assets and start a business immediately on AirBnb, RelayRides, TaskRabbit, or Zarrly.

Sell the by-products of being entrepreneurial. If you set up or design your own website, or optimize your own LinkedIn profile or Shopify store, there is a good chance there are others who would pay to have you do it for them. What might be easy for you to do, others might struggle with, or not want to bother with. There will always be a market of busy people who will pay for tasks.

Become a driver for: Lyft, Uber, or even Leap Transit.

Pet training, walking, servicing, grooming, sitters. Sign up on DogVaccy, or start your own business. If you can give pills and other basic medical treatments to cats, you can get work in most cities. Bonus points if you're familiar with birds or other exotic pets. Some people will pay for companionship for their animals while they are at work or away. Pets have become the black hole of American disposable income. If you love animals, you can make money.

Use technology to break into politics. These guys desperately need to be introduced to the 21st century, and—selfishly—I really want veterans to hold political office. I think most veterans would make better politicians than the average person obsessed with holding power for the wrong reasons. Use what you know that politicians don't, or offer them technology and social media services up front. Break into campaigns to learn, or start working for them immediately.

Coach a local sports team. Many high school teams pay money for this.

Teach your hobby. Paddle boarding, snowboarding, whatever it is, teach a class on what it is you love to do. Sell tickets via Eventbrite or PayPal. Ask your friends to buy tickets to your first event.

Physical Security: This can start in many different ways, from small jobs to protecting high-net-worth individuals' homes or families. There are also opportunities to teach personalized self-defense or security classes. Work for another company in this pursuit, start your own, or even teach a class about how to be safer on social media (for adolescents, young adults, seniors, etc.). Each age group has its own dangers: sexual predators, bullying, career embarrassment, financial scams. Research what the risks are and what trouble people have gotten into and then figure out how these problems could have been avoided. There are probably thousands of parents who would pay for information on how their kids can be safer and savvier online. Whether it's getting their kids to use CyberDust or another type of app which makes their movements less visible, this is what high-net-worth individuals will pay for. Why has nobody done Uber for security yet?!

Broadcast or podcast in whatever industry you're interested in. Interview every major player in the space. Do a podcast with video and upload it everywhere: YouTube, Vimeo, Soundcloud, iTunes, GPlay store, Amazon Store. If you can create a great show with information that people need, you'll be able to get advertisers or figure out a way to monetize it.

Master the Adobe Creative Suite. Whether it's audio or video production, masters of Adobe products can always find work. You can put advertisements for your work everywhere from Craigslist to Fiverr, oDesk, Elance, Micro lancer—you get the idea.

High End Designer: Same leads as above, except use sites like Dribble and 99U for leads. Get clients from large businesses in your local city.

Front End Developer: Master Ruby on Rails. There are many one-month to four-month-long classes where you can learn these skills.

Full-Stack Marketing. Learn everything about growth hacking. Find some online marketers and get started buying media for them. Branch out into other areas and learn everything about purchasing traffic, downloads, or acquiring customers for businesses. Metric based marketing, viral loops, growth engines, and low cost customer acquisition strategies are the skills of the marketers of the future.

Mobile Developer. Get started with iOS, and learn Appfolio to port Swift apps to Android.

Full-Stack Developer. Go to a boot camp, and then start doing freelance projects. Expand into an agency with services for clients, or build your own products.

Data Scientists. Master Excel, learn Hadoop, learn Java, become a master at manipulating data. Become friends with Wolfram Alpha. Freelance remotely or build a consultancy.

Home Care / Senior Care Health: The entire world is aging rapidly. Match up the best nurses and care providers with those who need them. Or, find a certification which will get you into this field as quickly as possible, and then build up a business from there. The global population is aging rapidly, and the number of diseases and ailments of an aging population means high demand for caretakers. Consider that by 2050, it's estimated that Alzheimer's will affect over 100 million people.[55]

Master a completely antifragile service skill such as bartending, waiting, singing, belly dancing, juggling. Spectacle? Yup. Silly? Yes. Trivial? Maybe, but will you always have a side stream of income? If you do it with just a bit more energy and aptitude than the average person, you absolutely will always have a job and opportunities. Will you always have options when the corporate employee with skills is laid off? Yes.

Go Forth and Conquer!

If taking action on any of these ideas were easy, everyone would do it. Most people won't do anything with information other then consume it. If you've made it this far I suspect you're not one of those people. Our country was built by those who built their entrepreneurial muscle through years of doing, learning, trying, testing, failing, and generally doing more than the average person. I hope these business ideas helped spark some of your own. Steal any or all of these ideas. Combine them with your own hustle, and best of luck as you take that drive to the frontline of capitalism.

Part 6: Rebuilding America

20

Vale

"It's easier to invent the future than to predict it."
–Alan Kay

"In the name of the best within you, do not sacrifice this world to those who are its worst. In the name of the values that keep you alive, do not let your vision of man be distorted by the ugly, the cowardly, the mindless in those who have never achieved his title. Do not lose your knowledge that man's proper estate is an upright posture, an intransigent mind and a step that travels unlimited roads. Do not let your fire go out, spark by irreplaceable spark, in the hopeless swamps of the approximate, the not-quite, the not-yet, the not-at-all. Do not let the hero in your soul perish, in lonely frustration for the life you deserved, but have never been able to reach. Check your road and the nature of your battle. The world you desired can be won, it exists, it is real, it is possible, it's yours."
–Ayn Rand

The Chances of a Lifetime

We're living in an amazing time in history. Everything we need or would like to learn is available at low cost via the Internet. Economic uncertainty might be crippling the developing world, but here in the U.S. we've remained largely unaffected. We have it pretty damn good as Americans, and the only things our country is suffering from right now are diseases of comfort. Now is a turning point in our country's history, and as veterans, we're perfectly trained to be the leaders that help rebuild America.

The culture of results, risks, and entrepreneurial spirit sparked by America's founders is re-emerging as we speak across the country. This culture of creating value and new technologies has been reborn in Silicon Valley and is now spreading all over America. This is not a culture that looks to others or the government for help or for handouts.

This is a culture of personal initiative and an obsession with tackling the biggest problems facing humanity. There has never been a better time to get in early and be a part of the development of amazing technologies. Whether it's a new career path, a startup, or a small business, now is the time to build skills and align ourselves with those rebuilding America. There isn't a better springboard into our own ventures than by immersing ourselves in the local startup community closest to us. Wherever there are communities innovating, veterans are needed.

We've stepped up and endured risks so that those here at home could take risks for what they believe. We've remained quiet while millions of people bemoaned conflicts abroad. As veterans, we've all faced the mental challenges of a military that's becoming increasingly bureaucratic, one where political correctness dominates. The ultimate meritocracy of fellow risk takers is waiting for us at home.

The future I see for America is incredibly bright. The future I see for any veteran who decides to control their destiny and capitalize on the valor they've already shown has no limit.

I look forward to the decades to come, because they will bring an ongoing stream of stories about veterans dominating in all walks of technology, entrepreneurship, and life. These stories will reaffirm what history has always taught those who listen; trauma and adversities always hold the master key to growth and accomplishment.

Veterans who refuse to accept the "disorder" part of PTS hold the power to reap the rewards of adversity. Those of us who look on mediocrity with scorn and aggressively seek knowledge will inevitably find success on our terms.

I'm Talking About Veterans Who:

- Make our own health the foundation of everything, and take full advantage of the preventative medicines, which include plenty of sleep, exercise, and a healthy diet.
- Insist on becoming CEOs in their relationships; Only deal with others seeking to be better; Recognize assholes, then work hard to distance themselves from them.
- Acquire skills in high demand
- Feel unapologetic as they layer skills on top of skills until they have become a monopoly of one and are paid ridiculously large sums of money for their expertise
- Relentlessly scour for secrets in their profession, tinkering at night and on the weekends to figure out how to create ways to do more with less, and forge alliances to turn the technology they create into viable businesses
- Align themselves with the culture coming out of Silicon Valley and seek to be a part of it, while helping to shape it, and mold it.
- Aren't afraid to join venture-backed startups where possible, or build the skills and business background necessary to launch their own technology
- Recognize that they have stood up and volunteered to put themselves at immense risk for a country which now needs to be lead by those risk-takers
- Desire, attain, control, and build the type of wealth and large families critical to the future of our country

To the veterans who choose this path or their own variation of it: I salute you, I encourage you, and I beg you to accept nothing less than victory for ourselves, and our families.

Our nation was founded by entrepreneurs turned venture capitalists. Our founders saw a tremendous opportunity to create the first democratic republic. The American democratic republic was an experiment with a completely revolutionary new form of government. It was a startup in many senses of the word. To borrow a term from computer science, the American republic was the first distributed system of government filled with checks and balances which recognized man's potential, but also man's lesser nature. America has proven to be the most successful experiment in the history of startups, government, and of all civilization. The founders of our country put everything on the line, in order for us to do the same.

The United States has brought staggering opportunity to the world.In 250 years, the fruits of America's entrepreneurs include:

Light: Tesla, Edison
Flight
Automobiles
Semi-Conductors
Computers
X-Rays
Solar Power, Wind Power
The Internet
UAV's and Drones
Nuclear Power
Biotechnology
Genomics: mapped the genome, etc.

Newer accomplishments include:

- A more antifragile Internet, built with a distributed system in mind. There are those working on creating this now.

- Bitcoin & the Blockchain: cryptocurrencies and the ability to transfer value while monitoring the transfer in a public way (in the form of bytes). The Internet of money. I realize that the pseudonym of the creator(s) of Bitcoin, Satoshi Nakamoto, is unknown and I want to keep it that way. But I'm betting Bitcoin was created based on innovations and investments made possible by America.
- Same day delivery on nearly anything we want, for a great price.
- Transportation via a few taps on a phone.
- The ability to buy and sell products and services much more easily. It's really a more liquid economy, although sophists insist on calling it the "sharing economy."
- Smartphones: More powerful than the computers of a decade ago. In 1993 dollars, many of us have a device worth millions of dollars at our disposal every day.
- Virtual Reality
- Robotics and AI
- Privatized Space Travel, much less expensive than what any government could create.
- A military which allows other countries to outsource their defense to us. We've also given the world an example of the first military in the history to defeat two global powers, Japan and Germany, and then pay to help them rebuild, trade with them, become allies, and help turn Japan into the third largest global economy.
- The defeat of the greatest evil of the 20th century: Communism (responsible for over 100 million deaths). Mainstream views and opinions might differ on this: did we defeat Communism, or did Communism defeat itself? I would argue that just our survival, along with the proliferation of jeans, rock'n roll, and other free spirited ideas, helped turn the tides toward freedom.
- Reinvention of the state. Our own government is currently in the process of being transformed and revitalized by technology.
- Montessori-style education. (AltSchool and your local Montessori style school)
- Silicon Valley Culture. Risk taking, flatter hierarchal structures,

personal responsibility, a religion of value creating through research, innovation, risk capital, and creative destruction.

- Cultural Alchemy. It used to just be a melting pot, but now we are starting to build a culture which welcomes all others, but still politely asks them to become American.
- The DARPA 100-year starship project. Hey, the government got ARPA Net right, maybe they'll get this one right too.
- The opportunity to make human life multi-planetary by colonizing Mars.

Earlier, we mentioned Lucius Seneca, the Roman Stoic, entrepreneur and venture capitalist. We've explored the Stoic concepts of apatheia and the practice of preparing to lose things in advance in order to place the proper value on them today. Seneca, and the rest of the Roman Stoics, would often end letters to their friends with the Latin word, "*Vale*." In English, *Vale* is most closely translated as "*be worthy*."

This was a poignant reminder which the Stoics would only make to true brothers and sisters. It marked true camaraderie in that it was meant to be a call to action to become something more. It recognized the friends' past accomplishments, yet encouraged them to always be worthy of continually greater feats and ventures.

As veterans, we must never let any of our brothers or sisters fall. We must fight to be worthy of the immense opportunities in front of us. With courageous and calculated risks, everything is within our grasp. And whatever we do, let's not "reintegrate." Let's always encourage one another to be remembered as the greatest generation of veterans. Let's be remembered as the ones who rebuilt America. *Vale.*

Glossary

A society can only improve so far as it's language allows. These terms and definitions below allow us to protect ourselves from the people who want our voice marginalized and diminished. I hope that this book prompts many to jump into high-growth fields, acquiring skills in such demand that you will become the entrepreneurs, investors, and leaders that will ensure America is restored. Maybe you've already noticed that people in entrepreneurial or business fields are already using these words in their common language, or maybe you haven't yet. But either way, there is a class of Americans already in the game, taking risks, and aggressively enjoying their lives to the fullest. These are the words and phrases those who have achieved apatheia are using on a daily basis.

America - The first definition in this book for a reason. The best country on earth, as long as we always seek to be and become better. This is the platform that currently contains companies and entrepreneurs taking aggressive bets to ensure humanity survives as a species.

Accelerator: A place and investment vehicle where companies start. They typically take a small equity stake of 5-7% of new businesses in exchange for connections, resources, and mentorship from entrepreneurs and investors who have worked directly in the space the individual hopes to create technology in. These are the new undergraduate and graduate programs for business. Business is best learned through doing, and accelerators are all about doing in the real world with real stakes and real incentives.

Accredited Investor: Someone who has a net worth (outside their primary residence) above a $1 million, or has had a household income of $300,000 or more over the last two years.

Apatheia: A mental state where nothing can be taken from you because you've striven to the best of your ability to properly value everything and everyone while you've interacted with them. This is a state where you don't fear future losses. You know they'll come, and you've recognized that the most prudent way to alleviate the sting of future loss lies in every present interaction with the people and things you love.

Angel Investor: An accredited investor that makes angel investments. Please beware of the huge number of "angels who aren't" or those individuals who are not accredited investors, but call themselves angel investors. Until they're accredited and you've seen proof, or Angel List has accredited them, stay away.

Angel List: The platform to find accredited investors, startups, private index funds to invest in. This is the social network of those who are building the future. A place for definite optimists. Indefinite optimists need not apply.

Antifragile: the opposite of fragile. A word used to define the interesting half of life, and one which gains from disorder. Coined by the former trader, rational flaneur, and philosopher Nassim Nicholas Taleb.

Evil: Any person or thing which seeks to make good actions, intent, and deeds feel impotent.

College: Part of an oligopoly, now largely devoted to fundamentalism(i.e the only hope for you lies in the hands of higher education and a credential).

Co-Working Space: A space where people work from different employers.

Compounding: The eighth wonder of the world. People think it only applies to money, missing out on the best part of life.

Flow States: Athletes describe this as being, "in the zone". Flow states can be achieved in work we love. Time dilates, our brain is flooded with chemicals, and we become in the zone, accomplishing a ton of things.

Future Proof: Blending the concept of antifragile with that of Silicon Valley and the amish to gain a mindset of how to select, use, avoid misuse, and generally help create technology in a way that ensures our own survival in a manner congruent with the survival of the rest of humanity.

Hipster: A walking hallmark and example of cultural rot. Refer to *Zero to One* for a breakdown.

Iatrogenics: Harm done by the person seeking to be the healer. Usually politicians, academics, doctors, or psychiatrists who are obsessed with treating the symptoms rather than recognizing the root causes of diseases, and who by treating the symptoms cause an entire stream of negative events.

Incubator: A home for startups, typically allowing them to work and get started with some type of equity arrangement or investment. Like a co-working space, except specifically for startups.

Interventionism: The obsession people have to do something, or intervene. Partly caused by a fear of others thinking they're not doing as much as they should. Political and bureaucratic systems at large organizations encourage this type of action.

Luddite: A detrimental term for a person opposed to new businesses that have created technology, or opposed to individuals who seek to create new technology.

Meditation / Mindfulness: Simply filtering thoughts from useful to useless, and building the practice to force your mind to only cling to thoughts which serve you. You don't have to sit down, hum, or grow your hair long to do this.

Neomania: An obsession with newness for the sake of newness.

Net Rate of Replacement: The number of children each parent needs to have in order to ensure a population grows. This number is currently 2.11 in the United States.

Perfect Technology: Technology that comes with virtuous stakes and incentives and limited downside. Print books would be an example of a perfect technology.

Private company: A company not publicly traded. Find some of the best ones in the portfolios of the top-tier venture capitalists we mentioned, along with those on Angel List.

Public markets: Oligopoly-manipulated equity markets such as NYSE, CME, etc,. Where insider trading is illegal for you, but not for others. You will never have an advantage, and there are only two strategies to make money: buy and hold for a really long time, or take advantage of suckers.

Rational Optionality: Preferring the ability to milk the option, or purchase the best option available, up to a certain point.

Reading: The gateway to becoming an autodidact. More powerful than just about any other form of instruction, but overlooked because people fail to do it in a way that benefits themselves.

Resistance: A term coined by Steven Pressfield for the force that immediately manifests anytime we seek to level up or become better. Also known as self-sabotage, excuses, fear, etc,.

Self-knowledge: Knowledge and work to figure out who you are, and what you want out of your life. Often can be found after coaching, therapy, reading, and bouts of strategic isolation.

Self-funding (also referred to as bootstrapping): I hate the term boot strapping—it just leaves a picture of idiocy in my mind—so if you're investing in yourself or your own company, in my mind that's self-funding, not strapping your boots. Unless you want to strap your boots and pull yourself up by those straps. Then go for it.

Sophist: A person who reasons speciously rather than soundly. One who prefers sophistication for sophistication's sake.

Startup Ecosystems: Anti-fundamentalist places where it's okay to create new things. These include startup communities that can be found amidst most major areas. These are places of cultural alchemy, science, measurement, creativity, etc. Dogmatism dies, data thrives.

Statist: A person who thinks government is the answer to all of their problems. Incredibly interested in stealing other people's money via taxes in order to spread it out to treat the symptoms of problems. Statists inherently create iatrogenics wherever they go, as they worship an organization that takes wealth and creates problems and unintended consequences that the virtuous in our society have to fight.

Tantalization / Tantalized Class: An entire group of people, largely in the middle class, who are comfortable but want more comfort and things, as opposed to leaving any lasting and meaningful legacy.

Technology: The ability to do more with less. Not always a good thing when it falls into the wrong hands. Requires virtue, self-knowledge, and long-term thinking to wield and directly appropriately.

Unlimited Upside: Exposure to huge opportunity. Usually comes in the form of equity or stock options. But publishing a great post on a site like LinkedIn has virtually unlimited upside and no real threats.

Vale: A mindset of always seeking to be worthy, of always improving and accepting defeat and losses on the road to victory.

Veteran: A class of around 2.8 million Americans, part of an all-volunteer military. Some of the last Americans willing to take personal risks with courage, who are comfortable operating with discomfort.

Special Thanks

To my parents who taught me to read, bought me countless books, and blessed me with a great childhood. You gave me some things I'm still trying to properly value: birth in America, teaching me empathy (it's all about those mirror neurons), and keeping me safe as a child. Mom and Dad, your support means the world to me, and I couldn't have done it without you. I love you!

To my wife, Stephanie, who sparked everything.

Shirley Grills: Grandma, you've supported and encouraged my writing ever since I started churning out articles for my school paper. You backed this Kickstarter project graciously, and I won't forget it! You and Grandpa have supported all my efforts, from joining the military, to writing this book. I love you and will make it up to Ohio to visit soon!

And now a special thanks to a few of the Backers that helped launch and backed the Kickstarter behind this book. You helped vindicate the project, and you're the reason it's now a reality. Thank you so much!

Major General (Ret), Alan Salisbury: Alan, it has been a pleasure to get to know you and volunteer at Code of Support. You're doing great work, and thank you for backing the project and reaching out to me! I look forward to attending and volunteering at more of the events Code of Support hosts, and sincerely appreciate your vote of confidence in me to write this book.

1SG (Ret), Richard Ruchty: When I got a message from Kickstarter notifying me that I had received my first large donation, I never thought it would be from a retired 1SG that I had never met, who lived in Hawaii! Receiving your donation was motivation and confirmation that this book was needed. Thank you for your support!

Nick Edwards: I appreciate you helping make this project possible. I was surprised, honored, and humbled by your donation. Thank you for your contribution, passion and funding that has helped so many Kickstarter projects succeed.

And a special thank you to fifteen extra special and highly motivated backers who made the Kickstarter for this book a reality:

Mark and Julie Henton
Bob and Brenda Delinger
Chuck and Melissa Schwaby
Tom and Kristy Smith
Kathy Reynolds
Jack Crouch
Sharron Scepkowski
Matthew King
Robert Gaskill
Austin Holmes
Jennifer and Brian Bain
Steve Keller
Mr. and Mrs. Atkinson
Terry David
Quentin Finney
Kathleen Hartmann

And last but not least, all those in Alpha Company, 1-175th Infantry whom I am honored to have served with. Thank you for passing on your wisdom, experience, and generally watching my back while I was in the military. I could not have asked for a better group of guys to train, and deploy with.

About the Author

Chad is an Army Infantry Veteran, entrepreneur, and author.

He has deployed to Iraq, Egypt, and provided security for the 56th Presidential Inauguration. You may have heard or seen him on the radio or T.V. as part of the military's 2011-2014 Defining Moments advertising campaign.

He's the author of several books including: *Future Proof*, and *Short Stories: Veterans After War*.

He works in technology, where he and his wife build and launch educational apps. They've built several products and apps, including College Majors, featured as *What's Hot* by Apple on iTunes. They're currently building a startup in the education space.

Speaking

I love speaking at events for veterans, businesses, students, and anyone who wants to get the most out of life. You can check out the places and events I've presented at, along with availability and booking information at:

www.ChadGrills.com/Speaking

Some of the topics I enjoy presenting on are listed below:

- Rebuilding America: The Entrepreneurial Leap
- Veterans: Rebuild America (my talk for Veterans and organizations that employ and work with them)
- HRSE: The Foundation of Future Leaders
- College or Not?

Connect with the Author

If you've made it this far, I'm honored and appreciative. Feel free to reach out, and if you've found any value in this book, I would love to hear about it.

I send out the occasional newsletter, with book updates, advanced copies of books, freebies, videos, and more. Feel free to sign up at: ChadGrills.com/subscribe

Or, feel free to tweet at me: @ChadGrills

Reviews Count

And now dear reader, you've finished the book, and we part ways.

If you're reading this on Kindle, when you turn this page, you'll be asked to review the book.

If you're reading this via paperback, feel free to leave a review at: http://www.amzn.com/B00NS7WISO

Reviews help this book become visible to others. No matter if it's short or long, I would love to hear what you think.

Thank you.

Notes

---

[1] Mriganka Sur, a neuroscientist, surgically rewired: M. Sur. 2003. How experience rewires the brain. Presentation at "Reprogramming the Human Brain" Conference, Center for Brain Health, University of Texas at Dallas, April 11.Doidge, Norman (2007-03-15). The Brain That Changes Itself: Stories of Personal Triumph from the Frontiers of Brain Science (Kindle Locations 4871-4873). Penguin Group US. Kindle Edition.

[2] Carducci, Bernardo J. (2009). "Basic Processes of Mischel's Cognitive-Affective Perspective: Delay of Gratification and Conditions of Behavioral Consistency". The Psychology of Personality: Viewpoints, Research, and Applications. John Wiley and Sons. pp. 443–4.

[3] http://cis.org/projecting-immigrations-impact-on-the-size-and-age-structure-of-the-21st-century-american-population

[4] http://fortune.com/2014/07/24/immigration-eb-5-visa-for-sale/

[5] http://www.slideshare.net/a16z/mobile-is-eating-the-world-40841467

[6] *A Whole New Mind*, Daniel Pink

[7] http://www.nytimes.com/2014/02/23/opinion/sunday/friedman-how-to-get-a-job-at-google.html?_r=1

[8] Many veterans are ambitious about politics and Shane Snow's book Smartcuts shows that our best presidents have been those who pursued multiple careers and did not spend too long in politics before they just jumped to the top (it's corrupting after all)- Consider Andrew Jackson who adamantly "paid his dues" and was promptly impeached. As Emerson said, "There are many things of which a wise man might wish to be ignorant." How politics has been done in the past is one of those things. Also, 22 veterans were just elected to congress and the senate in 2014. Veterans interested in politics can start emulating, working, or aligning themselves with this growing push of veteran leadership in Washington.

[9] Perlmutter, David (2013-09-17). Grain Brain: The Surprising Truth about Wheat, Carbs, and Sugar--Your Brain's Silent Killers (p. 63). Little, Brown and Company. Kindle Edition.

[10] Perlmutter, David (2013-09-17). Grain Brain: The Surprising Truth about Wheat, Carbs, and Sugar--Your Brain's Silent Killers (p. 10). Little, Brown and Company. Kindle Edition.

[11] Perlmutter, David (2013-09-17). Grain Brain: The Surprising Truth about Wheat, Carbs, and Sugar--Your Brain's Silent Killers (p. 85). Little, Brown and Company. Kindle Edition.

[12] Perlmutter, David (2013-09-17). Grain Brain: The Surprising Truth about

Wheat, Carbs, and Sugar--Your Brain's Silent Killers (p. 221). Little, Brown and Company. Kindle Edition.
[13] McGonigal, Ph.D., Kelly (2011-12-29). The Willpower Instinct: How Self-Control Works, Why It Matters, and What You Can Do To Get More of It (p. 9). Penguin Group. Kindle Edition.
[14] Ratey, John J. (2008-01-10). Spark: The Revolutionary New Science of Exercise and the Brain. Hachette Book Group. Kindle Edition.
[15] Doidge, Norman (2007-03-15). The Brain That Changes Itself: Stories of Personal Triumph from the Frontiers of Brain Science (p. 80). Penguin Group US. Kindle Edition.
[16] http://www.ncbi.nlm.nih.gov/pubmed/14871884
[17] http://www.sciencedaily.com/releases/2014/10/141028082508.htm
[18] http://www.brainpickings.org/2014/01/02/how-long-it-takes-to-form-a-new-habit/
[19] http://www.ncbi.nlm.nih.gov/pubmed/20643699
[20] http://www.maps.org/research/mdma/
[21] http://www.busybuildingthings.com/products/everyday-hustlin-calendar-2014
[22] https://fortune.com/2014/10/30/tony-robbins-best-advice-executive-coach/
[23] http://hbr.org/2004/01/what-makes-a-leader/ar/1
[24] http://hbr.org/2004/01/what-makes-a-leader/ar/1
[25] http://en.wikipedia.org/wiki/Tony_Hsieh
[26] https://www.youtube.com/watch?v=q6nphbmfaq8&list=uusyg0datjepo9lle-f_1l1g
[27] http://lowercasecapital.com/proprietor/
[28] https://www.youtube.com/watch?v=q6nphbmfaq8&list=uusyg0datjepo9lle-f_1l1g
[29] http://www.heritage.org/research/commentary/2009/12/why-the-stimulus-failed-to-create-jobs
[30] http://www.solarcity.com
[31] https://nextdoor.com
[32] https://www.uber.com/about
[33] http://www.shyp.com
[34] https://www.airbnb.com
[35] https://angel.co
[36] https://www.altschool.com
[37] http://bhpr.hrsa.gov/healthworkforce/supplydemand/usworkforce/primarycare/

[38] http://www.predilytics.com
[39] http://www.doctorondemand.com
[40] http://tricorder.xprize.org/teams
[41] http://en.wikipedia.org/wiki/IEX
[42] https://squareup.com
[43] http://a16z.com/portfolio/
[44] http://a16z.com/portfolio/
[45] http://a16z.com/portfolio/
[46] http://3drobotics.com/about-us/
[47] http://www.modrobotics.com
[48] http://www.spacex.com
[49] *Economic Facts and Fallacies*, Thomas Sowell pp. 117
[50] http://tradecrafted.com
[51] http://teamtreehouse.com
[52] http://expa.com/about
[53] http://1776dc.com
[54] http://www.trueventures.com
[55] Perlmutter, David (2013-09-17). Grain Brain: The Surprising Truth about Wheat, Carbs, and Sugar--Your Brain's Silent Killers (p. 31). Little, Brown and Company. Kindle Edition.

Made in the USA
Monee, IL
14 September 2026

62974510R00184